CAMBRIDGE LIBRARY COLLECTION

Books of enduring scholarly value

Darwin

Two hundred years after his birth and 150 years after the publication of 'On the Origin of Species', Charles Darwin and his theories are still the focus of worldwide attention. This series offers not only works by Darwin, but also the writings of his mentors in Cambridge and elsewhere, and a survey of the impassioned scientific, philosophical and theological debates sparked by his 'dangerous idea'.

On the Classification and Geographical Distribution of the Mammalia

English anatomist and biologist Richard Owen (1804-92), who in 1842 coined the word 'dinosaur', published this book in 1859, the year of On the Origin of Species. He collates ancient and recent studies of mammals in Western science before going on to present his own updated categorization of the class. Owen's eye for detail and range of scholarship are evident in this work, which is an extensive catalogue of mammals based on biological, geographical and anatomical characteristics. It incorporates, among other things, detailed classifications and sub-classifications of genus based on dental structures, food habits and cerebra. Owen's prose is lucid and precise and his investigations scrupulous, demonstrating the commitment that led him to become one of the foremost anatomists of his time. An appendix reveals Owen's rather conservative Idealist views on the hotly-debated theories of transmutation and extinction proposed by scientists such as Lamarck, Lyell and Darwin.

Cambridge University Press has long been a pioneer in the reissuing of out-of-print titles from its own backlist, producing digital reprints of books that are still sought after by scholars and students but could not be reprinted economically using traditional technology. The Cambridge Library Collection extends this activity to a wider range of books which are still of importance to researchers and professionals, either for the source material they contain, or as landmarks in the history of their academic discipline.

Drawing from the world-renowned collections in the Cambridge University Library, and guided by the advice of experts in each subject area, Cambridge University Press is using state-of-the-art scanning machines in its own Printing House to capture the content of each book selected for inclusion. The files are processed to give a consistently clear, crisp image, and the books finished to the high quality standard for which the Press is recognised around the world. The latest print-on-demand technology ensures that the books will remain available indefinitely, and that orders for single or multiple copies can quickly be supplied.

The Cambridge Library Collection will bring back to life books of enduring scholarly value across a wide range of disciplines in the humanities and social sciences and in science and technology.

On the Classification and Geographical Distribution of the Mammalia

Being the Lecture on Sir Robert Reade's Foundation, Delivered Before the University of Cambridge, in the Senate House, May 10, 1859. To which Is Added an Appendix 'On the Gorilla,' and 'On the Extinction and Transmutation of Species.'

RICHARD OWEN

CAMBRIDGE
UNIVERSITY PRESS

CAMBRIDGE UNIVERSITY PRESS

Cambridge New York Melbourne Madrid Cape Town Singapore São Paolo Delhi

Published in the United States of America by Cambridge University Press, New York

www.cambridge.org
Information on this title: www.cambridge.org/9781108001984

© in this compilation Cambridge University Press 2009

This edition first published 1859
This digitally printed version 2009

ISBN 978-1-108-00198-4

ON THE CLASSIFICATION

AND

GEOGRAPHICAL DISTRIBUTION

OF THE

MAMMALIA,

BEING THE

LECTURE ON SIR ROBERT READE'S FOUNDATION,

DELIVERED BEFORE THE

𝔘nibersity of Cambridge, in the Senate=House,

MAY 10, 1859.

TO WHICH IS ADDED AN APPENDIX

"ON THE GORILLA,"

AND

"ON THE EXTINCTION AND TRANSMUTATION OF SPECIES."

BY

RICHARD OWEN, F.R.S.

READE'S LECTURER IN THE UNIVERSITY OF CAMBRIDGE, SUPERINTENDENT OF THE NATURAL HISTORY DEPARTMENTS IN THE BRITISH MUSEUM, PRESIDENT OF THE BRITISH ASSOCIATION FOR THE ADVANCEMENT OF SCIENCE, FOREIGN MEMBER OF THE INSTITUTE OF FRANCE (ACADEMY OF SCIENCES), &c.

LONDON:

JOHN W. PARKER AND SON, WEST STRAND.

M.DCCC.LIX.

TO

THE REV. WILLIAM HENRY BATESON, D.D.,

MASTER OF ST JOHN'S COLLEGE,

AND

VICE-CHANCELLOR OF THE UNIVERSITY OF CAMBRIDGE.

My dear Sir,

I avail myself with pleasure of your permission to dedicate to you the present Discourse, which owes its existence principally to your favourable opinion of my ability to discharge the trust which you have done me the honour of confiding to me.

Believe me to be,

With the highest esteem and respect,

Your obliged and faithful Servant,

RICHARD OWEN.

MR VICE-CHANCELLOR AND GENTLEMEN OF THE
UNIVERSITY,

MY first impulse in availing myself of the privilege of
addressing you in this place, is, to give expression to
the deep sense which I entertain of the honour conferred
on me by my appointment to 'Sir Robert Reade's Lecture-
ship,' especially as it is the first which has been made since
the revival of that ancient foundation. Believe me, Sir, I truly
appreciate the favour of your choice, and am fully impressed
with the responsibilities which it involves. And if my ac-
knowledgments should seem curt or inadequate, I would be-
seech you to believe that this results from the wish not to tres-
pass too long on your most valuable time, but to devote to
the subject selected as much as may be of the period commonly allotted to an oral discourse.

In reviewing, for the choice of this subject, the field of
Natural Science in which I am a labourer, I desired to
select one that might be treated of with a certain degree
of completeness in a single Lecture, one that would enable
me to submit to you some of the more recent generalisations in Natural History, and at the same time exemplify
the applicability of that science, as a discipline, to the improvement of the intellect, and especially as a sharpener of the
faculties of observation and of methodical arrangement.

I trust that in the attempt to briefly unfold the Classification and Geographical Distribution of the *Mammalia* I
may attain the end I have in view.

The generalisation resulting in the idea of the natural
group of animals, so called, is one of ancient date. The
ZOOTOKA of Aristotle included the same outwardly diverse but
organically similar beings which constitute the MAMMALIA
of modern Naturalists. In that truly extraordinary compen-
dium of zoölogical and zootomical knowledge, the 'Περὶ
ζώων ἱστορίας[1],' animals generally, and by implication the

---

[1] Ed. Schneider, Leipzig, 1811, 4 Vols. 8vo.

B

Zootoka, or air-breathing vivipara, are divided according to the nature of their limbs into three sections:—1st, *Dipoda;* 2nd, *Tetrapoda ;* 3rd, *Apoda.* The first comprised the biped human race, the second the hairy quadrupeds, the third the whale-tribe, in which the limbs answering to the legs of man are wanting.

The second of these divisions, which includes the great majority of mammals, and is commonly regarded as the class itself, Aristotle subdivides into two great groups, according to the modification of the extremities. In the first group the foot is multifid, and a part of the digit—finger or toe—is left free for the exercise of the faculty of touch, the hard nail or claw being placed upon one side only; in the second group the digits are inclosed in hoofs: these groups are recognised in modern Zoology as the UNGUICULATA and UNGULATA.

Aristotle, in the generalised expressions of his observations on the various conditions of the teeth, has indicated subdivisions of the UNGUICULATA according to characters of the dental system. One subdivision includes those quadrupeds which have the front teeth trenchant, and the back teeth flattened, viz. the *Pithecoïda* or Ape-tribe. Another subdivision includes the quadrupeds with diversified· acuminated front teeth and interlocking serrated back teeth, viz. the *Karcharodonta*, or Carnivora; whilst the animals now known as ' Rodents' are indicated by a negative dental character.

With respect to the hoofed or Ungulate quadrupeds Aristotle in his generalisations on the organs of progressive motion divides them into *Dischidæ*, or bisulcate quadrupeds, and *Aschidæ*, or solidungulates, e. g. the horse and ass.

The term *Anepallacta*, by which Aristotle signified the animals in which the upper and lower teeth do not interlock, is applicable to the herbivorous quadrupeds generally ; in which the *Amphodonta*, or those with teeth in both jaws, e. g. the horse, are distinguished by him from those in which the front teeth are wanting in the upper jaw, e. g. the ox.

The bats were rightly recognised as true *Zootoka*, and the genus was defined as *Dermaptera*.

The apodal *Vivipara*, which form the third of Aristotle's more comprehensive groups, embraces the *Ketode*, now called *Cetacea*, and affords, by its position and co-ordinates in the great philosopher's zoological system, one of the most striking examples of his sagacity and research. In generalising, however, on modes of reproduction Aristotle includes certain sharks with the cetaceans, distinguishing the former by their gills, the latter by their blow-hole.

I ought, also, to remark that, although Aristotle has exemplified groups of animals which agree with many of the modern Classes, Orders, and Genera, their relative value is not so defined[1]; and his, in most respects, natural, assemblages would have commanded greater attention and been earlier and more generally recognised as the basis of later systems, had its immortal author more technically expressed an appreciation of the law of the subordination of characters; but Aristotle applies to each of his groups the same denomination, viz. γένος, genus; distinguishing, however, in some cases, the greater from the less.

Centuries elapsed ere any advance was made in the science of Zoology as it was bequeathed to the intellectual world by the mind of Aristotle. Of no other branch of human knowledge does the history so strongly exemplify the fearful phenomenon of the arrest of intellectual progress, resulting in the 'dark ages.' The well-lit torch which should have guided to further explorations of the mighty maze of animated nature was suffered to fall from the master-hand, and left to grow dim and smoulder through many generations ere it was resumed, fanned anew into brightness, and a clear view regained both of the extent of ancient discovery and of the right course to be pursued by modern research.

---

[1] See the just and discerning remarks on this subject by Dr WHEWELL, in his admirable *History of the Inductive Sciences*, 3rd ed., Vol. III. p. 289.

To JOHN RAY, an ornament of this University, I would
ascribe the merit of proposing a classification of the *Zootoka*,
which first claims attention as in any respect an advance
upon that taught by the Father of Natural History. It is
given in a tabular form in Ray's *Synopsis Methodica Anima-
lium Quadrupedum*, and is as follows:—(See p. 5).

In this Table the principle of the subordination of characters, or of their different values as applicable to groups of
different degrees of generalisation, is clearly exemplified; and
herein perhaps is its chief value. But, in the exclusion of
the *Dipoda* and *Apoda* of Aristotle, Ray manifests a less
philosophical appreciation of the extent and essential nature
of the class *Zootoka* than his great predecessor. He is also
inferior in the discernment of the real significance of certain
modifications of zoological characters. Aristotle was not de-
ceived either by the claw-like shape of the hoofs of the camel,
or by the degree of subdivision of those of the elephant; he
knew that both quadrupeds were, nevertheless, essentially
Ungulate[1].

LINNÆUS first definitely and formally restored the great
natural class I am now treating of to its Aristotelian inte-
grity; and, applying to it that happy instinct of discernment
of significant outward characters which had enabled him to
effect so much for the sister Science of Botany, he proposed
for it the name MAMMALIA.

The active cultivation of the science of observation stimulated by Ray, Linnæus and Buffon, had brought to light
instances, e. g. in certain lizards, of viviparous quadrupeds
which differ in structures of classific importance from the
Zootoka tetrapoda of Aristotle. Certain forms of true fishes
were now known to bring forth their young alive, as well as
the fish-like *Ketode*. The term *Zootoka* ceased to be applicable, exclusively, to the class of which Aristotle had sketched
out the bounds; and Naturalists gladly accepted and have
since retained the neat and appropriate and truly distinctive

[1] 'Καὶ ἀντὶ ὀνύχων χηλὰς ἔχει.'

A TABLE OF VIVIPAROUS FOUR-FOOTED ANIMALS.

Viviparous hairy animals or quadrupeds are,—

Ungulate, and these either

Solidipedous, as the HORSE, ASS, ZEBRA.

Bisulcate, which are

Ruminants with horns, that are

Persistent, as in the Ox, SHEEP, GOAT,

or

Deciduous, as in the STAG.

or

Not Ruminants, as the HOG.

Quadrisulcate, as the RHINOCEROS, HIPPOPOTAMUS.

Unguiculate, whose feet are either

Bifid, as in the CAMEL, or

Multifid, which are

With *digits* adhering together, and covered with a common integument, so that the extremities alone are visible at the margin of the foot, and are covered with obtuse nails, as in the ELEPHANT.

With *digits* in some measure distinct and separable from each other, the nails being

Depressed, as in APES,

or

Compressed, where the incisor teeth are

Many, in which group all the animals are carnivorous and rapacious, or at least insectivorous, or subsist on insects with vegetable matter:

The larger ones with the

Muzzle short, and head rounded, as the Feline tribe ; or with the

Muzzle long, as the Canine tribe ;

The smaller ones with a long slender body, and short extremities, as the Weasel or Vermine[1] tribe ;

or

Two very large, of which tribe all the species are phytivorous, as the HARE.

1 Genus *Vermineum*, from their worm-like form.

term proposed by Linnæus,—the term which was suggested by the outward and visible part of that apparatus by which the warm-blooded viviparous animals exclusively nourish their new-born young[1].

Linnæus, like Ray, founds his primary divisions of the class MAMMALIA on the locomotive organs; but his secondary divisions or orders are taken chiefly from modifications of the dentary system. The following is an abridged scheme of his arrangement[2]:—

<table>
<tr><td rowspan="7">MAMMALIA.</td><td rowspan="4">Unguiculate</td><td>Front teeth, none in either jaw . . .</td><td>BRUTA.</td></tr>
<tr><td>Front teeth, cutters 2, laniaries 0 . . .</td><td>GLIRES.</td></tr>
<tr><td>Front teeth, cutters 4, laniaries 1 . .</td><td>PRIMATES.</td></tr>
<tr><td>Front teeth, piercers (6, 2, 10), laniaries 1</td><td>FERÆ.</td></tr>
<tr><td rowspan="2">Ungulate</td><td>Front teeth, in both upper and lower jaw.</td><td>BELLUÆ.</td></tr>
<tr><td>Front teeth, none in the upper jaw . .</td><td>PECORA.</td></tr>
<tr><td>Muticate</td><td>Teeth variable . . . .</td><td>CETE.</td></tr>
</table>

On comparing the three preceding systems, it will be found that the most important errors of arrangement have been committed, not by Aristotle, but by the modern naturalists. Both Ray and Linnæus have mistaken the character of the horny parts enveloping the toes of the elephant, which do not defend the upper part merely, as is the case with claws, but embrace the under parts also, forming a complete case or hoof.

With respect to Linnæus, however, it must be observed, that although he has followed Ray in placing the elephant in the unguiculate group of quadrupeds, he has not overlooked the great natural divisions which the latter naturalist adopted from Aristotle; and his *Ungulata* is the more natural in the degree in which it approaches the corresponding group in the Aristotelian system.

I now proceed to the arrangement of the Mammalia proposed by CUVIER in the last edition of his classical work entitled '*Le Règne Animal distribué d'après son organisation.*'

Adopting the same threefold primary division of the class MAMMALIA as his predecessors, CUVIER subdivides it into

---

[1] Aristotle knew that the Cetacea were mammiferous: ‘τὰ’ (δὲ δύο μὲν μαστούς) ‘δ’ ἐντὸς, ὥσπερ δελφίς.’

[2] From the *Systema Naturæ*, ed. XII. Holmiæ, Tom. I. p. 24.

more naturally defined orders, according to various characters afforded by the dental, osseous, generative and locomotive systems, which his great anatomical knowledge had made known to him.

That heterogeneous order which Linnæus—prepossessed in favour of the easily recognisable outward character by which he distinguished the class—had characterised by the '*Mammæ pectorales binæ:* dentes primores incisores : superiores IV *paralleli*[1],' was shewn, by the correlation of anatomical distinctions with the threefold modification of the limbs of the *Primates*, to be divisible into as many distinct orders. The hands on the upper limbs alone, and the lower limbs destined to sustain the trunk erect, characterised the order *Bimana*, the equivalent of the Linnæan genus *Homo*. The genus *Simia* of Linnæus, with hands on the four extremities, became the order *Quadrumana* of Cuvier. The genus *Vespertilio* with the 'manus palmatæ volitantes' formed the group *Cheiroptera*, answerable to the *Dermaptera* of Aristotle.

RAY had pointed out certain viviparous quadrupeds with a multifid foot as being "anomalous species," instancing as such "the tamandua, the armadillo, the sloth, the mole, the shrew, the hedgehog, and the bat." The first three species are associated with the scaly ant-eaters (*Manis*) of Asia and Africa, with the Australian spiny ant-eaters (*Echidna*), and with the more strange duck-moles (*Ornithorhynchus*) of the same part of the world, to form the order *Edentata* of Cuvier, which answers to that called *Bruta* by Linnæus, if the elephant and walrus be removed from it. The rest of *Ray's* anomalous species exemplify the families *Cheiroptera* and *Insectivora* of the Cuvierian system, in which they are associated with the true *Carnivora* in an order called 'Carnassiers,' answering to the *Feræ* of Linnæus.

Cuvier had early noticed the relation of the Australian pouched mammals, as a small collateral series, to the

---

[1] Tom. cit. p. 24.

unguiculate mammals of the rest of the world; 'some,' he writes, 'corresponding with the *Carnivora*, some with the *Rodentia*, and others again with the *Edentata*, by their teeth and the nature of their food.' They formed a family of the *Carnassiers* in the first edition of the '*Règne Animal*[1],' but were raised to the rank of an order under the name *Marsupialia* in the second edition, where they terminate that series of the *Unguiculata*, which possess the three kinds of teeth—incisors, canines and molars.

The hoofed animals (UNGULATA, 'animaux à sabots') are binarily divided into those that do, and those that do not, chew the cud; the former constituting the order *Pachydermata*, the latter that of *Ruminantia*.

The third primary group or subclass of Mammalia is indicated, but without receiving any name distinct from that of the single order *Cetacea* exemplifying it in the Cuvierian system—an order which would be equivalent to the *Mutica* of the Linnæan system, save that the manatee which Linnæus placed in the same group as the elephant is associated with the whale in the *Règne Animal*.

The Mammalian system of CUVIER is exemplified in the subjoined Table:—(See p. 9).

Important as was the improvement it presented on previous classifications, the progress of anatomical and physiological knowledge, mainly stimulated by the writings and example of Cuvier himself, soon began to make felt the defects of his system. Shortly after its proposition, the zoological mind began to be disagreeably impressed by the results of the application of the characters employed by Cuvier in the formation of the primary and secondary groups of the class; the sloth, for example, being placed above the horse, the mole above the lynx, and the bat above the dog: even the *Ornithorhynchus paradoxus*—shewn by accurate anatomical scrutiny to be the most reptilian of the mammalian class—takes

---

[1] 8vo., 1816.

9

TABLE OF THE SUBCLASSES AND ORDERS OF THE MAMMALIA, ACCORDING TO CUVIER.

CLASS	SUBCLASS	ORDER	FAMILY OR GENUS	EXAMPLE
MAMMALIA	UNGUICULATA	BIMANA	Homo	Man.
		QUADRUMANA	Catarrhina	Ape.
			Platyrrhina	Marmoset.
			Strepsirrhina	Lemur.
		CARNARIA [1] (With three kinds of teeth.)	Cheiroptera	Bat.
			Insectivora	Hedgehog. Shrew. Mole.
			Carnivora	Bear. Dog. Seal.
		MARSUPIALIA	Didelphys	Opossum.
			Phalangista	Phalanger.
			Macropus	Kangaroo.
			Phascolomys	Wombat.
		RODENTIA (Without canines.)	Claviculata	Rat.
			Non-claviculata	Hare.
		EDENTATA (Without incisors.)	Bradypus	Sloth.
			Dasypus	Armadillo.
			Myrmecophaga	Anteater.
			Monotremata	Echidna. Ornithorhynchus.
	UNGULATA	PACHYDERMATA	Proboscidia	Elephant.
			Ordinaria	Hog. Tapir. Horse.
		RUMINANTIA	Solidungula	Sheep.
	MUTILATA	CETACEA	Herbivora	Dugong.
			Ordinaria	Whale.

1 Written *Carnassiers* by Cuvier.

precedence of the colossal and sagacious elephant in the Cuvierian scheme[1].

The profound admiration and respect which I have always
entertained for my chief instructor in Zootomy and Zoology,
never blinded me to the necessity of much modification of his
arrangement of the *Mammalia*. The question, more especially, of the truly natural and equivalent primary groups of
the class, has been present to my mind whenever I have been
engaged in dissecting the rarer forms which have died at the
Zoological Gardens in London, or on other occasions. But I
propose first to submit to you, as briefly and clearly as I am
able, the results of this store of anatomical knowledge as applicable to the true organic characters of the class MAMMALIA.

Mammals are distinguished outwardly by an entire or
partial covering of hair[2], and by having teats or mammæ—
whence the name of the class.

All mammals possess mammary glands and suckle their
young: the embryo or fœtus is developed in a womb. Their
leading anatomical character is, the highly vascular and mi-

Fig. 1.

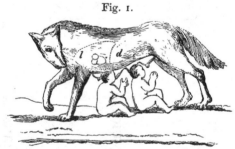

nutely cellular structure of the lungs, (fig. 1, *l*,) which are
freely suspended in a thoracic cavity separated by a musculo-
tendinous partition or 'diaphragm' from the abdomen, (ib. *d*.)

---

[1] The modifications consequently proposed by Geoffroy St Hilaire, Illiger,
De Blainville, C. L. Bonaparte, J. E. Gray, Waterhouse, Milne Edwards,
Lesson, Wagner, Nilsson, Oken, Macleay, Sir E. Home, Gervais, and others,
have been cited and commented upon in my Papers communicated to the Linnæan Society (*Proceedings*, 1857) and the Geological Society (*Proceedings*, Nov.
1847, pp. 135—140).

[2] The fœtal Cetacea shew tufts of hair on the muzzle.

Mammals, like Birds, have a heart composed of two ventricles and two auricles, and have warm blood: they breathe quickly; but inspiration is performed chiefly by the agency of the diaphragm; and the inspired air acts only on the capillaries of the pulmonary circulation.

The blood-discs are smaller than in Reptiles, and, save in the Camel-tribe, are circular in form. The right auriculo-ventricular valve is membranous, and the aorta bends over the left bronchial tube.

The kidneys are relatively smaller and present a more compact figure than in the other vertebrate classes; their parenchyma is divided into a cortical and medullary portion, and the secreting tubuli terminate in a dilatation of the excretory duct, called the pelvis: they derive the material of their secretion from the arterial system. Their veins are simple, commencing by minute capillaries in the parenchyma and terminating generally by a single trunk on each side in the abdominal vena cava: they never anastomose with the mesenteric veins.

The liver is generally divided into a greater number of lobes than in Birds. The portal system is formed by veins derived exclusively from the spleen and chylopoietic viscera. The cystic duct, when it exists, always joins the hepatic, and does not enter the duodenum separately. The pancreatic duct is commonly single.

The mouth is closed by soft flexible muscular lips: the upper jaw is composed of palatine, maxillary and premaxillary bones, and is fixed; the lower jaw consists of two side-halves, or rami, which are simple or formed by one bony piece, and are articulated by a convex (fig. 3, b)⁻ or flat condyle to the base of the zygomatic process, and not to the tympanic element, of the temporal bone; the base of the coronoid process (ib. c) generally extends along the space between the condyloid and the alveolar processes. The jaws of Mammals with few exceptions are provided with teeth, which are arranged in a single row; they are always lodged in sockets, and never anchylosed with the substance of the jaw. The tongue is fleshy, well-developed, with the apex more or less free. The posterior nares are protected by a soft palate, and the larynx by an epiglottis: the rings of the trachea are generally cartilaginous and incomplete behind: there is no inferior larynx. The œsophagus is continued without partial dilatations to the stomach, which varies in its structure according to the nature of the food, or the quantity of nutriment to be extracted therefrom.

The trunk-vertebræ of Mammalia have their bodies ossified from three centres, and present for a longer or shorter period of life a discoid epiphysis at each extremity. They are articulated by concentric ligaments with interposed glairy fluid forming what are called the intervertebral substances; the articulating surfaces are generally flattened, but sometimes, as in the necks of certain Ruminants, they are concave behind and convex in front. The cervical vertebræ are seven in number, neither more nor less. The lumbar vertebræ are more constant and usually more numerous than in other classes of vertebrate animals. The atlas is articulated by concave articular processes to two convex condyles, which are developed from the ex-occipital elements, or neurapophyses, of the last cranial vertebra. The tympanic element of the temporal bone is restricted in function to the service of the organ of hearing, and never enters into the articulation of the lower jaw. The olfactory nerves escape from the cranial cavity through numerous foramina of a cribriform plate. The optic foramina are always distinct from one another.

The scapula is generally an expanded plate of bone; the coracoid, with two (monotrematous) exceptions, appears as a small process of the scapula. The sternum consists of a narrow and usually simple series of bones: the sternal portions of the ribs are generally cartilaginous and fixed to the vertebral portions without the interposition of a distinct articulation: there are no gristly or bony abdominal ribs or abdominal sternum. The pubic and ischial arches are generally complete, and united together by bony confluence on the sternal aspect, so that the interspace of the two pelvic arches is converted into two holes, called 'foramina obturatoria.'

The sclerotic coat of the eye is a fibrous membrane, and never contains bony plates. In the quantity of aqueous humour and the convexity of the lens Mammals are generally intermediate between Birds and Fishes. The organ of hearing is characterized by the full development of the cochlea with a lamina spiralis: there are three distinct ossicles in the tympanum; the membrana tympani is generally concave externally; the meatus auditorius externus often commences with a complicated external ear, having a distinct cartilaginous basis. The external apertures of the organ of smell are provided with moveable cartilages and muscles, and the extent of the internal organ is increased by accessory cavities or sinuses which communicate with the passages including the turbinated bones.

There are few characters of the osseous system common, and at the same time peculiar, to the class Mammalia. The following may be cited:—

1. Each half or ramus of the mandible consists of one bony piece developed from a single centre: the condyle is convex or flat, never concave. This has proved a valuable character in the determination of fossils.

2. The second or distal bone, called 'squamosal,' in the 'zygomatic' bar continued backward from the maxillary arch, is not only expanded, but is applied to the side-wall of the cranium, and developes the articular surface for the mandible, which surface is either concave or flat.

3. The presphenoid is developed from a centre distinct from that of the basisphenoid.

In no other class of vertebrate animals are these osteological characters present.

The cancellous texture of mammalian bone is of a finer and more delicate structure than in Reptiles, and forms a closer network than in Birds. The microscopic radiating cells are relatively smaller and approach more nearly to the spheroid form.

The Mammalia, like *Reptilia* and *Pisces*, include a few genera and species that are devoid of teeth; the true ant-eaters (*Myrmecophaga*), the scaly anteaters or pangolins (*Manis*), and the spiny monotrematous anteater (*Echidna*), are examples of strictly edentulous Mammals. The Orni-thorhynchus has horny teeth, and the whales (*Balæna* and *Balænoptera*) have transitory embryonic calcified teeth, suc-ceeded by whalebone substitutes in the upper jaw. The female Narwhal seems to be edentulous, but has the germs of two tusks in the substance of the upper jaw-bones; one of these becomes developed into a large and conspicuous weapon in the male Narwhal, whence the name of its genus *Monodon*.

The examples of excessive number of teeth are presented, in the order *Bruta*, by the priodont Armadillo, which has ninety-eight teeth: and in the Cetaceous order by the Cacha-lot, which has upwards of sixty teeth, though most of them are confined to the lower jaw; by the common Porpoise, which has between eighty and ninety teeth: by the Gangetic Dol-phin, which has one hundred and twenty teeth; and by the true Dolphins (*Delphinus*), which have from one hundred to one hundred and ninety teeth, yielding the maximum number in the class Mammalia.

When the teeth are in excessive number, as in the Arma-dillos and Dolphins above cited, they are small, equal, or sub-equal, and usually of a simple conical form.

In most other mammals particular teeth have special forms for special uses; thus, the front teeth, (figs. 2 and 3, *i*,) from being commonly adapted to effect the first coarse division of the food, have been called cutters or *incisors;* and the back teeth, (ib. *m*,) which complete its comminution, grinders or *molars;* large conical pointed teeth situated behind the in-cisors, and adapted, by being nearer the insertion of the biting muscles, to act with greater force, are called holders, tearers, laniaries, or more commonly *canines*, (ib. *c,*) from being well developed in the Dog and other Carnivora.

It is peculiar to the class Mammalia to have teeth implanted in sockets by two or more fangs; but this can only happen to teeth of limited growth, and generally characterizes the molars and premolars: perpetually growing teeth require the base to be kept simple and widely excavated for the persistent pulp. In no mammiferous animal does anchylosis of the tooth with the jaw constitute a normal mode of attachment. Each tooth has its peculiar socket, to which it firmly adheres by the close co-adaptation of their opposed surfaces, and by the firm adhesion of the alveolar periosteum to the organized cement which invests the fang or fangs of the tooth.

True teeth implanted in sockets are confined, in the Mammalian class, to the maxillary, premaxillary, and mandibular or lower maxillary bones, and form a single row in each. They may project only from the premaxillary bones, as in the Narwhal; or only from the lower maxillary bone, as in *Ziphius;* or be limited to the superior and inferior maxillaries and not present in the premaxillaries, as in the true *Ruminantia* and most *Bruta* (Sloths, Armadillos, Orycteropes). In most Mammals teeth are situated in all the bones above mentioned.

The teeth of the Mammalia usually consist of hard unvascular dentine, defended at the crown by an investment of enamel, and everywhere surrounded by a coat of cement.

The coronal cement is of extreme tenuity in Man, Quadrumana and the terrestrial Carnivora; it is thicker in the Herbivora, especially in the complex grinders of the Elephant.

Vertical folds of enamel and cement penetrate the crown of the tooth in the ruminating and many other Ungulata, and in most Rodents, characterizing by their various forms the genera of those orders.

No Mammal has more than two sets of teeth. In some

species the tooth-matrix does not develope the germ of a
second tooth, destined to succeed the one into which the matrix
has been converted; such a tooth, therefore, when completed
and worn down, is not replaced. The Sperm Whales, Dolphins, and Porpoises are limited to this simple provision of
teeth. In the Armadillos and Sloths, the want of generative
power, as it may be called, in the matrix is compensated by
the persistence of the matrix, and by the uninterrupted growth
of the teeth.

In most other Mammalia, the matrix of the first-developed
tooth gives origin to the germ of a second tooth, which sometimes displaces the first, sometimes takes its place by the side
of the tooth from which it has originated.

All those teeth which are displaced by their progeny are
called 'temporary,' deciduous, or milk-teeth, (figs. 2 and 3,
d, 1...4); the mode and direction in which they are displaced
and succeeded, viz. from above downwards in the upper, from
below upwards in the lower, jaw, in both jaws vertically—are
the same as in the Crocodile; but the process is never re-
peated more than once in any mammalian animal. A con-
siderable proportion of the dental series is thus changed; the
second or 'permanent' teeth having a size and form as suitable
to the jaws of the adult, as the 'temporary' teeth were adapted
to those of the young animal.

Those permanent teeth, which assume places not previously occupied by deciduous ones, are always the most posterior in their position, and generally the most complex in
their form. The term 'molar' or 'true molar' is restricted
to these teeth (fig. 2 and 3, m). The teeth between them and
the canines are called 'premolars,' (ib. p); they push out the
milk-teeth, (ib. d,) and are usually of smaller size and simpler
form than the true molars.

Thus the class Mammalia, in regard to the times of formation and the succession of the teeth, may be divided into
two groups, viz. *Monophyodonts*[1] or those that generate a

[1] μόνος, once; φύω, I generate; ὁδούς, tooth.

single set of teeth, and the *Diphyodonts*[1] or those that generate two sets of teeth. But this dental character is not so associated with other organic characters as to indicate natural or equivalent sub-classes.

In the Mammalian orders with two sets of teeth, these organs acquire individual characters, receive special denominations, and can be determinated from species to species. This differentiation of the teeth is significative of the high grade of organization of the animals manifesting it.

Originally, indeed, the names 'incisors,' 'canines,' and 'molars,' were given to the teeth, in Man and certain Mammals, as in Reptiles and Fishes, in reference merely to the shape and offices indicated by those names; but they are now used as arbitrary signs, in a more fixed and determinate sense. In some Carnivora, *e. g.*, the front teeth have broad tuberculate summits adapted for nipping and bruising, while the principal back-teeth are shaped for cutting and work upon each other like the blades of scissors. The front-teeth in the Elephant project from the upper jaw, in the form, size and direction of long pointed horns. Indeed, shape and size are the least constant of dental characters in the Mammalia; and the homologous teeth are determined, like other parts, by their relative position, by their connexions, and by their development.

Those teeth which are implanted in the premaxillary bones, and in the corresponding part of the lower jaw, are called 'incisors' (fig. 2, *i*), whatever be their shape or size. The tooth in the maxillary bone, which is situated at or near to the suture with the premaxillary, is the 'canine,' as is also that tooth in the lower jaw (ib. *c*), which, in opposing it, passes in front of the upper one's crown when the mouth is closed. The other teeth of the first set are the 'deciduous molars' (*d.* 1—3); the teeth which displace and succeed them vertically are the 'premolars' (*p.* 1—3); the more posterior

---

[1] δὶς, twice; φύω and ὀδούς. See *Philosophical Transactions*, 1850, p. 493.

teeth, which are not displaced by vertical successors, are the 'molars' properly so called (*m.* 1—4).

Fig. 2.

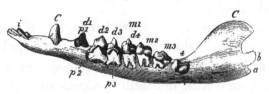

Lower Jaw of a young Opossum (*Didelphys*).

I have been led, chiefly by the state of the dentition in most of the early forms of both carnivorous and herbivorous Mammalia, which flourished during the eocene tertiary periods, to regard 3 incisors, 1 canine, and 7 succeeding teeth, on each side of both jaws, as the type formula of diphyodont dentition.

Three of the seven teeth may be 'premolars' (fig. 2, *p.* 1—3), and four may be true 'molars' (ib. *m.* 1—4); or there may be four premolars (fig. 3, *p.* 1—4), and three true molars (ib. *m.* 1—3). This difference forms a character of an

Fig. 3.

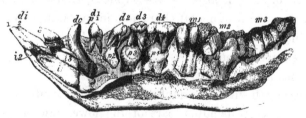

Lower Jaw of a young Pig (*Sus.*)

ordinal group in the mammalian class[1]. The essential nature of the distinction is as follows: true molars (ib. *m.*) are a backward continuation of the first series of teeth (ib. *d.*); they are developed in the same primary groove of the fœtal gum; they are 'permanent' because they are not pushed out by the successional teeth (ib. *p.*), called 'dents de remplacement' by Cuvier. Seven teeth developed in the primary groove is,

[1] *Outlines of a Classification of the Mammalia, Trans. Zool. Soc.* Vol. II. p. 330 (1839).

therefore, the typical number of first teeth, beyond the canines. If, as in *Didelphys* (fig. 2), the anterior three develope tooth-germs which come to perfection in a 'secondary groove,' there are then 3 deciduous teeth, 3 premolars, and 4 true molars: if, as in *Sus*, fig. 3, the anterior four of the 'primary' teeth develope tooth-germs, which grow in a secondary groove, there are then 4 deciduous teeth, 4 premolars, and 3 true molars. The first true molar of the marsupial (fig. 2, *m.* 1, *d.* 4), is thus seen to be the homologue of the last milk-molar of the placental (fig. 3, *d.* 4).

The Hog, the Mole, the Gymnure and the Opossum, are among the few existing quadrupeds which retain the typical number and kinds of teeth. In a young Hog of ten months (fig. 3), the first premolar, *p.* 1, and the first molar, *m.* 1, are in place and use together with the three deciduous molars, *d.* 2, *d.* 3, and *d.* 4; the second molar, *m.* 2, has just begun to cut the gum; *p.* 2, *p.* 3, and *p.* 4, together with *m.* 3, are more or less incomplete, and will be found concealed in their closed alveoli[1].

The last deciduous molar, *d.*4, has the same relative superiority of size to *d.* 3 and *d.* 2, which *m.* 3 bears to *m.* 2 and *m.* 1; and the crowns of *p.* 3 and *p.* 4 are of a more simple form than those of the milk-teeth, which they are destined to succeed. When the milk-teeth are shed, and the permanent ones are all in place, their kinds are indicated, in the genus *Sus*, by the following formula :—

$$i. \frac{3-3}{3-3}, \quad c. \frac{1-1}{1-1}, \quad p. \frac{4-4}{4-4}, \quad m. \frac{3-3}{3-3} = 44;$$

which signifies that there are on each side of both upper and lower jaws 3 incisors, 1 canine, 4 premolars, and 3 molars, making in all 44 teeth, each tooth being distinguished by its appropriate symbol, viz. *p.*1 to *p.*4, *m.*1 to *m.*3. This number of teeth is never surpassed in the placental diphyodont series.

[1] I recommend this easily acquired 'subject' to the young zoologist for a demonstration of the most instructive peculiarities of the mammalian dentition. He will see that the premolars must displace deciduous molars in order to rise into place : the molars have no such relations.

When the premolars and the molars are below this typical number, the absent teeth are missing from the back part of the molar series, and usually from the fore part of the premolar series. The most constant teeth are the fourth premolar and the first true molar. These being known by their order and mode of development, the homologies of the remaining molars and premolars are determined by *counting the molars from before backwards*, e. g. 'one,' 'two,' 'three,' and the premolars *from behind forwards*, 'four,' 'three,' 'two,' 'one.' The incisors are counted from the median line, commonly the foremost part, of both upper and lower jaws, outwards and backwards. The first incisor of the right side is the homotype, transversely, of the contiguous incisor of the left side in the same jaw, and vertically, of its opposing tooth in the opposite jaw; and so with regard to the canines, premolars, and molars; just as the right arm is the homotype of the left arm in its own segment, and also of the right leg of a succeeding segment. It suffices, therefore, to reckon and name the teeth of one side of either jaw in a species with the typical number and kinds of teeth, e. g. the first, second, and third incisors,—the first, second, third, and fourth premolars, —the first, second, and third molars; and of one side of both jaws in any case.

I have been induced to dwell thus long on the dental characters of the class *Mammalia*, because they have not been rightly defined in any systematic or elementary work on zoology, although an accurate formula and notation of the teeth are of more use and value in characterizing genera in this than in any other class of animals.

Mammals may be surpassed in the rapidity with which the blood circulates, in the extent and completeness of the respiratory processes, in bodily temperature, in the concomitant vigour of the muscular actions; all which superiorities, in Birds, for example, result in those marvellous powers of flight with which the feathered class is privileged. But in their psychical phenomena the *Mammalia*, as a class, excel all other animals. Let me exemplify this by reference to the reproductive economy in the vertebrate series.

The instinctive sense of dependence upon another, manifested by the impulse to seek out a mate,—which impulse, even in fishes, is sometimes so irresistible that they throw themselves on shore in the pursuit,—this first step in the supercession of the lower and more general law of individual- or self-preservation, although not first introduced at the vertebrate stage of the animal series, is never departed from after that stage has been gained. To this sexual relation is next added a self-sacrificing impulse of a higher kind, viz. the parental instinct. As we rise in the survey of vertebrate phenomena, we see the entire devotion of self to offspring in the patient incubation of the bird, in the unwearied exertions of the Swift or the Hawk to obtain food for their callow brood when hatched; in the bold demonstration which the Hen, at other times so timid, will make to repel threatened attacks against her cowering young.

Still closer becomes the link between the parent and offspring in the Mammalian class, by the substitution, for the exclusion of a passive irresponsive ovum, of the birth of a living young, making instinctive irresistible appeal, as soon as born, to maternal sympathy; deriving nutriment immediately from the mother's body, and both giving and receiving pleasure by that act.

These beautiful foreshadowings of higher attributes are, however, transitory in the brute creation, and the relations cease, as soon as the young quadruped can provide for itself. Preservation of offspring has been superinduced on self-preservation, but there is as yet no self-improvement: this is the peculiar attribute of mankind. The human species is characterised by the prolonged dependence of a slowly maturing offspring on parental cares and affections, in which are laid the foundations of the social system, and time given for instilling those principles on which Man's best wisdom and truest happiness are based, and by which he is prepared for another and a higher sphere of existence. In this destination alone may we discern an adequate end and purpose in the great organic scheme developed upon our planet.

The progressive gradations in this scheme will be further exemplified as I proceed to explain the principles and characters by which I have been guided in the formation of the primary groups or divisions of the class *Mammalia*.

Prior to the year 1836 it was held by comparative anatomists that the brain in *Mammalia* differed from that in all other vertebrate animals by the presence of the large mass of transverse white fibres, called 'corpus callosum' by the anthropotomist; which fibres, overarching the ventricles and diverging as they penetrate the substance of either hemisphere of the cerebrum, bring every convolution of the one into communication with those of the other hemisphere, whence the other name of this part—the 'great commissure.' In that year I discovered that the brain of the kangaroo, the wombat, and some other marsupial quadrupeds, wanted the 'great commissure;' and that the cerebral hemispheres were connected together, as in birds, only by the 'fornix' and 'anterior commissure[1].' Soon afterward, I had the opportunity of determining that the same deficiency of structure prevailed in the *Ornithorhynchus* and *Echidna*[2].

As many other modifications of structure, more or less akin to those characterizing birds and reptiles, were found to be associated with the above oviparous type of brain, together with some remarkable peculiarities in the economy of reproduction[3], I pointed out that the *Mammalia* might be divided into 'placental' and 'implacental[4].'

Impressed, however, with the fact that such binary division, like that which might be based upon the leading differences of dentition, was too unequal to be natural, the larger

---

[1] See *Philosophical Trans.* for 1837, p. 87.
[2] Art. MONOTREMATA, *Cyclopædia of Anatomy and Physiology*, Vol. III. p.383.
[3] Art. MARSUPIALIA, tom. cit. p. 257.
[4] Art. MAMMALIA, tom. cit. p. 244.

group never presenting the same degree of correspondence of organic structure as the smaller moiety, I continued to pursue investigations, with the view of gaining an insight into the more natural and equivalent primary groups of the *Mammalia;* having my attention more especially directed to the cerebral organ in this quest.

In 1842, I was able to demonstrate, in the 'Hunterian Course of Lectures' delivered at the Royal College of Surgeons, the leading modifications of the mammalian brain, and their peculiar value in classification by reason of their association with concurrent modifications of other systems of organs.

Nevertheless there were genera of Mammals, *e. g.* the sloths, anteaters, armadillos, roussettes, giraffes, rhinoceroses, &c. to which the cerebral test had to be applied. Fortunately the rare species of these genera successively arrived at the Zoological Gardens in London, and afforded me the means of applying that test; so that, at length, having dissected the brain in one species at least, of almost every genus or natural family of the Mammalian class, I felt myself in a position to submit to the judgment of my fellow-labourers in zoology, at the Linnæan Society, in 1857, the generalised results of such dissections, comprising a fourfold primary division of the MAMMALIA, based upon the four leading modifications of cerebral structure in that class.

In some mammals the cerebral hemispheres are but feebly and partially connected together by the 'fornix' and 'anterior commissure:' in the rest of the class the part called 'corpus callosum' is added, which completes the connecting or 'commissural' apparatus.

With the absence of this great superadded commissure[1] is associated a remarkable modification of the mode of development of the offspring, which involves many other modifications; amongst which are the presence of the bones called 'marsupial,' and the non-development of the deciduous body

---

[1] *On the Structure of the Brain in Marsupial Animals, Philos. Trans.* 1837.

concerned in the nourishment of the progeny before birth, called 'placenta;' the young in all this 'implacental' division being brought forth prematurely, as compared with the rest of the class.

This first and lowest primary group, or subclass, of Mammalia is termed, from its cerebral character, *LYENCEPHALA*[1], —signifying the comparatively loose or disconnected state of the cerebral hemispheres. The size of these hemispheres (fig. 4, A) is so small that they leave exposed the olfactory ganglions (*a*), the cerebellum (C), and more or less of the optic lobes (B); their surface is generally smooth; the anfractuosities, when present, are few and simple.

Fig. 5.

Fig. 4.

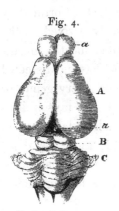

Brain of Opossum.

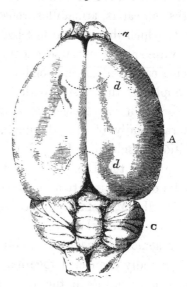

Brain of Beaver.

The next well marked stage in the development of the brain is where the corpus callosum (indicated in fig. 5, by the dotted lines *d, d*) is present, but connects cerebral hemispheres as little advanced in bulk or outward character as in the preceding subclass; the cerebrum (A) leaving both the olfactory lobes (*a*) and cerebellum (C) exposed, and being commonly

[1] λύω, to loose; ἐγκέφαλος, brain.

smooth, or with few and simple convolutions in a very small proportion, composed of the largest members, of the group. The mammals so characterized constitute the subclass *LISSEN-CEPHALA*[1] (fig. 5).

The third leading modification of the Mammalian cerebrum is such an increase in its relative size, that it extends over more or less of the cerebellum; and generally more or less over the olfactory lobes. Save in very few exceptional cases of the smaller and inferior forms of *Quadrumana* (fig. 6), the superficies is folded into more or less numerous gyri or convolutions (fig. 7),—whence the name *GYRENCEPHALA*, which I propose for the third subclass of Mammalia[2].

Fig. 7.

Fig. 6.

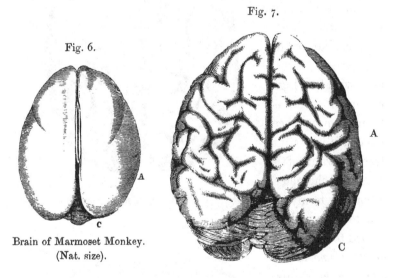

Brain of Marmoset Monkey.
(Nat. size).

Brain of Chimpanzee.
(Half nat. size).

In Man the brain presents an ascensive step in development, higher and more strongly marked than that by which the preceding subclass was distinguished from the one below it. Not only do the cerebral hemispheres overlap the olfactory lobes and cerebellum, but they extend in advance of the

[1] λισσός, smooth ; ἐγκέφαλος, brain.
[2] γυρόω, to wind about ; ἐγκέφαλος, brain.

one and further back than the other (figs. 8 & 9). Their posterior development is so marked that anthropotomists have assigned to that part the character and name of a 'third lobe:' it is peculiar and common to the genus *Homo*: equally peculiar is the 'posterior horn of the lateral ventricle' and the 'hippocampus minor,' which characterize the hind lobe of each hemisphere. The superficial grey matter of the cerebrum, through the number and depth of the convolutions, attains its maximum of extent in Man.

Peculiar mental powers are associated with this highest form of brain, and their consequences wonderfully illustrate the value of the cerebral character; according to my estimate of which, I am led to regard the genus *Homo* as not merely a representative of a distinct order, but of a distinct subclass, of the Mammalia, for which I propose the name of *Archencephala* (fig. 9)[1].

Fig. 8.

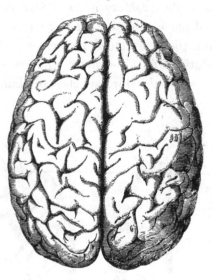

Brain of Negro, upper view.

Fig. 9.

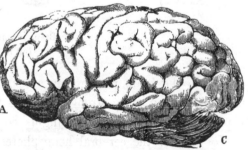

Ib. Side view, one-third nat. size.

---

[1] ἄρχω, to overrule ; ἐγκέφαλος, brain.

With this preliminary definition of the organic characters, which appear to guide to a conception of the most natural primary groups of the class MAMMALIA, I next proceed to define the groups of secondary importance, or the subdivisions of the foregoing subclasses.

The Lyencephalous Mammalia are unguiculate : some have the 'optic lobes' simple, others partly subdivided, or complicated by accessory ganglions, the lobes being then called 'bigeminal bodies.'

The *LYENCEPHALA* with simple optic lobes are 'edentulous' or without calcified teeth, and are devoid of external ears, scrotum, nipples, and marsupial pouch : they are true 'testiconda :' they have a coracoid bone extending from the scapula to the sternum, and also an epicoracoid and episternum as in Lizards : they are unguiculate and pentadactyle, with a supplementary tarsal bone supporting a perforated spur in the male. The order so characterized is called 'MONOTREMATA,' in reference to the single excretory and generative outlet, which, however, is by no means peculiar to them among Mammalia. It includes two genera—*Echidna* and *Ornithorhynchus.* Of the first, the species are terrestrial, insectivorous, chiefly myrmecophagous, having the beak-like slender jaws, and long cylindrical tongue of the true anteaters ; but they are covered, like the hedgehog, with spines. Of the second genus, the species are aquatic, with a flattened beak, like that of a duck, which is used in the anserine manner to extract insects and worms from the mud : but they are clothed with a close fine fur like that of a mole, whence the name 'duck-mole' by which these anomalous quadrupeds are commonly known to the colonists. Both genera of Monotremes are strictly limited to Australia and Tasmania.

The *LYENCEPHELA* with divided optic lobes, forming the 'corpora bigemina' and 'quadrigemina' of anthropotomists, have teeth, and with rare exceptions, the three kinds, viz. incisors, canines, and molars. They are called MARSUPIA-LIA, because they are distinguished by a peculiar pouch or duplicature of the abdominal integument, which in the males is everted, forming a pendulous bag, and in the females is inverted, forming a hidden pouch containing the nipples and usually sheltering the young for a certain period after their birth: they have the marsupial bones in common with the Monotremes; a much varied dentition, especially as regards the number of incisors, but usually including 4 true molars; and never more than 3 premolars[1] (fig. 2): the angle of the lower jaw (ib. a) is more or less inverted[2].

With the exception of one genus, *Didelphys*, which is American, and another genus *Cuscus*, which is Malayan, all the known existing Marsupials belong to Australia, Tasmania, and New Guinea. The grazing and browsing Kangaroos are rarely seen abroad in full daylight, save in dark rainy weather. Most of the Marsupialia are nocturnal. Zoological wanderers in Australia, viewing its plains and scanning its scrubs by broad daylight, are struck by the seeming absence of mammalian life; but during the brief twilight and dawn, or by the light of the moon, numerous forms are seen to emerge from their hiding-places and illustrate the variety of marsupial life with which many parts of the continent abound. We may associate with their low position in the mammalian scale the prevalent habit amongst the Marsupialia of limiting the exercise of the faculties of active life to the period when they are shielded by the obscurity of night.

The premature birth of the offspring, and its transference to the tegumentary pouch, in which it remains suspended to the nipple for a period answering to that of uterine life in higher mammals, relate to the peculiarities of the climate of Australia.

The adventurous and much-enduring explorers of that continent bear uniform testimony to the want of water as the

[1] *Outlines of a Classification of the Marsupialia*, Trans. Zool. Soc. Vol. II. 1839.
[2] For other Osteological and Dental characteristics of the Marsupialia, see the paper above cited, and that On the Osteology of the Marsupialia, Trans. Zool. Soc. Vol. II. p. 379 (1838).

chief cause of their sufferings and danger. During the dry season the rivers are converted into pools, 'few and far between;' and the drought is sometimes continued so long as to dry up these. An ordinary non-marsupial quadruped, such as the wild cat or fox, having deposited her young in the nest or burrow, would in such a climate, at the droughtiest period of her existence, be compelled to travel a hundred, perhaps two hundred miles, in order to quench her thirst. Before she could return her blind and helpless litter would have perished. By the marsupial modification the mother is enabled to carry her offspring with her in the long migrations necessitated by the scarcity of water.

With the climatal peculiarities of Australia, therefore, we may connect the peculiar modifications of those members of the mammalian class which are most widely distributed over that continent. But the principle of final causes receives more especial illustrations from the contingent particulars of the marsupial organization. The new-born Kangaroo is an inch in length, naked, blind, with very rudimental limbs and tail: in one which I examined the morning after the birth, I could discern no act of sucking: it hung, like a germ, from the end of the long nipple, and seemed unable to draw sustenance therefrom by its own efforts. The mother, accordingly, is provided with a peculiar adaptation of a muscle (cremaster) to the mammary gland, by which she can inject the milk from the nipple into the mouth of the pendulous embryo. Were the larynx of the little creature like that of the parent, the milk might—probably would—enter the windpipe and cause suffocation: but the fœtal larynx is cone-shaped, with the opening at the apex, which projects, as in the whale-tribe, into the back aperture of the nostrils, where it is closely embraced by the muscles of the 'soft palate.' The air-passage is thus completely separated from the fauces, and the injected milk passes in a divided stream on either side the base of the larynx into the œsophagus. These correlated modifications of maternal and fœtal structures, designed with especial reference to the peculiar conditions of both mother and off-spring, afford, as it seems to me, irrefragable evidence of Creative foresight.

The LISSENCEPHALA, or smooth-brained placental Mammalia, form a group which I consider as equivalent to the LYENCEPHALA or Implacentals; and which includes the following orders, *Rodentia, Insectivora, Cheiroptera* and *Bruta.*

The RODENTIA are characterized by two large and long curved incisors in each jaw, separated by a wide interval from the molars; the teeth being so constructed, and the jaw so articulated, as to effect the reduction of the food to small particles by acts of rapid and continued gnawing, whence the name of the order. The orbits are not separated from the temporal fossæ. The male glands pass periodically from the abdomen into a temporary scrotum, and are associated with prostatic and vesicular glands. The placenta is commonly discoid, but is sometimes a circular mass (Cavy), or flattened and divided into three or more lobes (Lepus). The Beaver and Capybara are the giants of the order, which chiefly consists of small, numerous, prolific and diversified unguiculate genera, subsisting wholly or in part on vegetable food. Some Rodents, *e. g.* the Lemmings, perform remarkable migrations, the impulse to which, unchecked by dangers or any surmountable obstacles, seems to be mechanical. Many Rodents build very artificial nests, and a few manifest their constructive instinct in association. In all these inferior psychical manifestations we are reminded of Birds. Many Rodents hibernate like Reptiles. They are distributed over all continents. About two-thirds of the known species of *Mammalia* belong to the Rodent order.

The transition from the Marsupials to the Rodents is made by the Wombats; and a transition from the Marsupials is made, by an equally easy step, through the smaller Opossums to the INSECTIVORA. This term is given to the order of small smooth-brained Mammals, the molar teeth of which are bristled with cusps, and are associated with canines and incisors: they are unguiculate, plantigrade, and pentadactyle, and they have complete clavicles. Like Rodents, they are temporary testiconda, and have large prostatic and vesicular glands: like most other *Lissencephala,* the Insectivora have a discoid or cup-shaped placenta. They do not exist in South America and Australia; their office in these continents is fulfilled by Marsupialia; but true Insectivora abound in all the other continents and their contiguous islands.

The order CHEIROPTERA, with the exception of the modification of their digits for supporting the large webs that serve as wings, repeat the chief characters of the Insectivora: a few, however, of the larger species are frugivorous and have corresponding modifications of the teeth and stomach. The mammæ are pectoral in position.

The most remarkable examples of periodically torpid Mammals are to be found in the terrestrial and volant Insectivora. The frugivorous Bats differ much in dentition from the true Cheiroptera, and would seem to conduct through the Colugos or Flying Lemurs, directly to the Quadrumanous order. The Cheiroptera are cosmopolitan.

The order BRUTA, called *Edentata* by Cuvier, includes two genera (*Myrmecophaga* and *Manis*) which are devoid of teeth; the rest possess those organs, which, however, have no true enamel, are never displaced by a second series, and are very rarely implanted in the premaxillary bones. All the species have very long and strong claws. The ischium as well as the ilium unites with the sacrum; the orbit is not divided from the temporal fossa. The Three-toed Sloths (*Bradypus*) manifest their affinity to the oviparous Vertebrata by the supernumerary cervical vertebræ supporting false ribs and by the convolution of the wind-pipe in the thorax; and I may add that the unusual number—three and twenty pairs—of ribs, forming a very long dorsal, with a short lumbar, region of the spine, in the Two-toed Sloth (*Cholœpus*), recalls a lacertine structure. The same tendency to an inferior type is shown by the abdominal testes, the single cloacal outlet, the low cerebral development, the absence of medullary canals in the long bones in the Sloths, and by the great tenacity of life and long-enduring irritability of the muscular fibre, in both the Sloths and Anteaters[1].

The order Bruta is but scantily represented at the present period. One genus, *Manis* or Pangolin, is common to Asia and Africa; the *Orycteropus* is peculiar to South Africa; the rest of the order, consisting of the genera *Myrmecophaga*, or true Anteaters, *Dasypus* or Armadillos, and *Bradypus* or Sloths, are confined to South America.

Having defined the orders or subdivisions of the two foregoing subclasses, I may remark that the LYENCEPHALA cannot be regarded as equivalent merely to one of the orders, say *Rodentia*, of the LISSENCEPHALA, without undervaluing the anatomical characters which are so remarkable and distinct in the marsupial and monotrematous animals. The anatomical peculiarities of the edentulous LYENCEPHALA[2] appear to me to be, at least, of ordinal importance. In these deductions I hold the mean between those who, with Geoffroy St Hilaire, would make a distinct class of the *Monotremata*, and those who, with Cuvier, would make the Monotremes a mere family of the *Edentata*. In like manner, whilst I regard the LYENCEPHALA as forming a group of higher rank than an order, I do not consider it as forming an equivalent primary group to that formed by all the placental Mammalia.

The true value of the LYENCEPHALA is that of one of four

---

[1] This latter vital character attracted the notice of the earliest observers of these animals. Thus Marcgrave and Piso narrate of the Sloth :— 'Cor motum suum validissimè retinebat, postquam exemptum erat e corpore per semihorium :—exempto corde cæteris visceribus multo post se movebat et pedes lentè contrahebat sicut dormituriens solet.' Buffon, who quotes the above from the *Historia Naturalis Brasiliæ*, p. 322, well remarks, ' Par ces rapports, ce quadrupède se rapproche non seulement de la tortue, dont il a la lenteur, mais encore des autres reptiles et de tous ceux qui n'ont pas un centre du sentiment unique et bien distinct.'—*Hist. Naturelle*, 4to, Tom. XIII. p. 45.

[2] See my article *Monotremata*, in the *Cyclopædia of Anatomy*, part xxvi. 1841.

primary divisions or subclasses of the Mammalia; its true equivalency is with the *Lissencephala*, and all its analogical relations are to be found more directly in that smooth-brained subclass than in the Placentalia at large.

The following Table exemplifies the correspondence of the groups in the Lyencephalous and Lissencephalous series:—

LYENCEPHALA.	LISSENCEPHALA.
Rhizophaga[1]	Burrowing *Rodentia*.
Poëphaga[1]	*Dipodidæ* and *Leporidæ*.
Petaurus	*Pteromys*.
Phalangistidæ	*Sciuridæ* and prehensile-tailed arboreal Rodents.
Phascolarctos	*Bradypus*.
Perameles and *Myrmecobius*.	*Erinaceidæ*.
Chœropus	*Macroscelis*.
Didelphys and *Phascogale* .	*Soricidæ*.
Dasyuridæ	*Centetes, Gymnura*.
Echidna	*Manis*.

Besides the more general characters by which the *Lissencephala*, in common with the *Lyencephala*, resemble Birds and Reptiles, there are many other remarkable indications of their affinity to the Oviparous Vertebrata in particular orders or genera of the subclass. Such, e. g., are the cloaca, convoluted trachea, supernumerary cervical vertebræ and their floating ribs, in the three-toed Sloth; the numerous trunk-ribs in the two-toed Sloth; the irritability of the muscular fibre, and persistence of contractile power in the Sloths and some other Bruta; the long, slender, beak-like edentulous jaws and gizzard of the Anteaters; the imbricated scales of the equally edentulous Pangolins, which have both gizzard and gastric glands like the proventricular ones in birds; the dermal bony armour of the Armadillos like that of loricated Saurians; the quills of the Porcupine and Hedgehog; the brilliant iridescent colours of the fur of the Cape-mole (*Chrysochlora aurea*); the proventriculus of the Dormouse and Beaver; the pre-

---

[1] *On the Classification of the Marsupialia, Trans. of the Zool. Soc.* Vol. II. p. 315, 1839.

valence of disproportionate development of the hind limbs in
the *Rodentia*, coupled, in the Jerboa, with confluence of the
three chief metatarsals into one bone, as in birds; the keeled
sternum, and wings of the Bats; the aptitude of the *Cheiroptera*, *Insectivora*, and certain *Rodentia* to fall, like Reptiles,
into a state of true torpidity, associated with a corresponding
faculty of the heart to circulate carbonized or black blood:—
these, and the like indications of coaffinity with the LYEN-
CEPHALA to the Oviparous air-breathing Vertebrata, have
mainly prevailed with me against an acquiescence in the
elevation of different groups of the LISSENCEPHALA to a higher
place in the Mammalian series, and in their respective association, through some single character, with better-brained orders,
according to Mammalogical systems which, at different times,
have been proposed by zoologists of deserved reputation.
Such, e. g., as the association of the long-clawed *Bruta* with
the *Ungulata*[1], and of the shorter-clawed Shrews, Moles and
Hedgehogs, as well as the Bats, with the *Carnivora*[2]; of the
Sloths with the *Quadrumana*[3]; of the Bats with the same
high order[4]; and of the *Insectivora* and *Rodentia* in immediate
sequence after the Linnean 'Primates,' as in the latest published 'System of Mammalogy,' from a distinguished French
author[5].

So far as their ordinal affinities are known, the most
ancient Mammals, the fossil remains of which have been found
in secondary strata, are either ly- or liss-encephalous, and belong
either to the *Marsupialia* or the *Insectivora*. (Appendix A).

In the GYRENCEPHALA we look in vain for those marks of
affinity to the oviparous vertebrate animals which have been in-
stanced in the preceding subclasses; although, it is true, that
when we proceed to consider the subdivisions of the GYREN-

---

[1] Macleay, *Linn. Trans.* Vol. XVI. (1833); Gray, Dr. J. E., *Mammalia in
the British Museum*, 12mo, 1843, p. xii.

[2] Cuvier, *Règne Animal*, 1829, p. 110.

[3] De Blainville, *Ostéographie*, 4to, Fasc. I. p. 47 (1839).

[4] Linnæus, *Systema Naturæ*, Ed. 12, Tom. I. p. 26.

[5] Prof. Gervais, *Zoologie et Paléontologie Français*, 4to, 1852, p. 194.

CEPHALA, we seem at first to descend in the scale by finding in that wave-brained subclass a group of animals, having the form of Fishes: but a high grade of mammalian organization is masked beneath this form.

The GYRENCEPHALA are primarily subdivided, according to modifications of the locomotive organs, into three series, for which the Linnean terms may well be retained; viz. *Mutilata*, *Ungulata* and *Unguiculata*, the maimed, the hoofed, and the clawed series.

These limb-characters can only be rightly applied to the gyrencephalous subclass; they do not indicate natural groups, save in that section of the Mammalia. To associate the LYENCEPHALA and LISSENCEPHALA with the unguiculate GYREN-CEPHALA into one great primary group, as in the Mammalian systems of Ray, Linnæus and Cuvier, is a misapplication of a solitary character akin to that which would have founded a primary division on the discoid placenta or the diphyodont dentition. No one has proposed to associate the unguiculate Bird or Lizard with the unguiculate Ape; and it is but a little less violation of natural affinities to associate the Mono-tremes with the Quadrumanes in the same primary (unguicu-late) division of the Mammalian class.

The three primary divisions of the GYRENCEPHALA are of higher value than the ordinal divisions of the LISSENCEPHALA; just as those orders are of higher value than the representative families of the LYENCEPHALA.

The *Mutilata*, or the maimed Mammals with folded brains, are so called because their hind limbs seem, as it were, to have been amputated; they possess only the pectoral pair of limbs, and these in the form of fins: the hind end of the trunk expands into a broad, horizontally flattened, caudal fin. They have large brains with many and deep convolutions, are naked, and have neither neck, scrotum, nor external ears.

The first order, called CETACEA, in this division are either edentulous or monophyodont, and the latter have teeth of one kind and usually of simple form. They are 'testiconda,' and have no 'vesiculæ seminales.' The mammæ are pudendal; the placenta is diffused; the external nostrils—single or double—are on the top of the head, and called spiracles or 'blow-holes.' They are marine, and, for the most part, range the unfathomable ocean; though with certain geographical limits as respects species. The 'right whale' of the northern hemisphere (*Balæna mysticetus*) is represented by a distinct species (*Balæna australis*) in the southern hemisphere: the high temperature of the waters at the equatorial zone bars the migration of either from one pole to the other. True Cetacea feed on fishes or marine animals.

The second order, called SIRENIA, have teeth of different kinds, incisors which are preceded by milk-teeth, and molars with flattened or ridged crowns, adapted for vegetable food. The nostrils are two, situated at the upper part of the snout; the lips are beset with stiff bristles; the mammæ are pectoral; they are 'testiconda,' but have 'vesiculæ seminales.' The Sirenia exist near coasts or ascend large rivers; browsing on fuci, water plants, or the grass of the shore. There is much in the organization of this order that indicates its nearer affinity to members of the succeeding division, than to the cetaceous order.

The Dugongs (*Halicore*) inhabit the Red sea, the Malayan Archipelago, and the soundings of the Australian coasts: the Manatees (*Manatus*) frequent the shores of tropical America and Africa.

In the *Ungulata* the four limbs are present, but that portion of the toe which touches the ground is incased in a hoof, which blunts its sensibility and deprives the foot of prehensile power. With the limbs restricted to support and locomotion, the Ungulates have no clavicles; the two bones of the fore leg are fixed together in the position anatomists call 'prone;' as a general rule hoofed quadrupeds feed on vegetables.

A particular order, or suborder, of this group is indicated by fossil remains of certain South American genera, e. g. *Toxodon* and *Nesodon*, with long, curved, rootless teeth, having a partial investment of enamel, and with certain peculiarities of cranial structure: the name TOXODONTIA is proposed for this order, all the representatives of which are extinct[1].

A second remarkable order, most of the members of which have also passed away, is characterized by two incisors in the form of long tusks; in one genus (*Dinotherium*) projecting from the under jaw, in another genus (*Elephas*) from the upper jaw, and in some of the species of a third genus (*Mastodon*) from both jaws. There are no canines: the molars are few, large and transversely ridged; the ridges sometimes few and mammillate, often numerous and with every intermediate gradation. The nose is prolonged into a cylindrical trunk, flexible in all directions, highly sensitive, and terminated by a prehensile appendage like a finger: from this peculiar organ is derived the name PROBOSCIDIA given to the order. The feet are pentadactyle, but the toes are indicated only by divisions of the hoof; the placenta is annular; the mammæ are pectoral.

Elephants are dependent chiefly upon trees for food. One species now finds the conditions of its existence in the rich forests of tropical Asia; a second species in those of tropical Africa. Why, we may ask, should not a third be living at the expense of the still more luxuriant vegetation watered by the Oronoko, the Essequibo, the Amazon, and the La Plata, in tropical America? Geology tells us that at least two kinds of Elephant (*Mastodon Andium* and *Mast. Humboldtii*) formerly did derive their subsistence, along with the great Megatherioid beasts, from that abundant source: two other kinds of Elephant (*Mastodon ohioticus* and *Elephas texianus*) existed in the warm and temperate latitudes of North America. Twice as many species of Mastodon and Elephant, distinct from all the others, roamed in pliocene times in the same latitudes of Europe. At a later or pleistocene period, a huge elephant, clothed with wool and hair, obtained its food from hardy trees, such as now grow in the 65th degree of north latitude; and

[1] *Philosophical Transactions*, 1853, p. 291.

abundant remains of this *Elephas primigenius* (as it has been prematurely called, since it was the last of our British elephants) have been found in temperate and high northern latitudes in Europe, Asia and America.   This, like other Arctic animals, was peculiar in its family for its range in longitude. The Musk Buffalo was its contemporary in England and Europe, and still lingers in the northernmost parts of America.

I have received evidences of Elephantine species from China and Australia, proving the proboscidian pachyderms to have once been the most cosmopolitan of hoofed herbivorous quadrupeds.

Both the proboscidian and toxodontal orders of UNGULATA may be called aberrant : the dentition of the latter, and several particulars of the organization of the Elephant, indicate an affinity to the *Rodentia ;* the cranium of the Toxodon, like that of the Dinothere, resembles that of the *Sirenia* in its remarkable modifications.

The typical Ungulate quadrupeds are divided, according to the odd or even number of the toes, into PERISSODACTYLA and ARTIODACTYLA[1]: the single hoof of the horse, the triple hoof of the tapir, exemplify the first: the double hoof of the camel, the quadruple hoof of the hippopotamus, exemplify the second.   In the perissodactyle or odd-toed UNGULATA, the dorso-lumbar vertebræ differ in number in different species, but are never fewer than twenty-two; the femur has a third trochanter, and the medullary artery does not penetrate the fore part of its shaft.   The fore part of the astragalus is divided into two very unequal facets.   The os magnum and the digitus medius which it supports are large, in some disproportionately so, and the digit is symmetrical : the same applies to the ectocuneiform and the digit which it supports in the hind foot.   If the species be horned, as the Rhinoceros, the horn is single ; or, if there be two, they are placed on the median line of the head, one behind the other, each being thus an odd horn.

[1] From περισσοδάκτυλος, qui digitos habet impares numero ; and ἄρτιος, par, δάκτυλος, digitus.—*Quarterly Journal of the Geological Society*, No. 14, May, 1848.

There is a well-developed post-tympanic process which is separated by the true mastoid from the paroccipital in the Horse, but unites with the lower part of the paroccipital in the Tapir, and seems to take the place of the mastoid in the Rhinoceros and Hyrax. The hinder half, or a larger proportion, of the palatines enters into the formation of the posterior nares, the oblique aperture of which commences in advance either of the last molar, or, as in most, of the penultimate one. The pterygoid process has a broad and thick base and is perforated lengthwise by the ectocarotid. The crown of from one to three of the hinder premolars is as complete as those of the molars : that of the last lower milk-molar is commonly bilobed. To these osteological and dental characters may be added some important modifications of internal structure, as, e. g., the simple form of the stomach and the capacious and sacculated cæcum, which equally evince the mutual affinities of the odd-toed or perissodactyle quadrupeds with hoofs, and their claims to be regarded as a natural group of the *Ungulata*. Many extinct genera, e. g. *Coryphodon, Pliolophus, Lophiodon, Tapirotherium, Palæotherium, Ancitherium, Hipparion, Acerotherium, Elasmotherium*, &c., have been discovered, which once linked together the now broken series of Perissodactyles, represented by the existing genera *Rhinoceros, Hyrax, Tapirus*, and *Equus*. The placenta is replaced by a diffused vascular villosity of the chorion in all the recent genera of this order, excepting the little *Hyrax*, in which there is a localised annular placenta, as in the Elephant. But the diffused placenta occurs in some genera of the next group, shewing the inapplicability of that character to exact classification.

In the even-toed or ' artiodactyle' Ungulates, the dorsolumbar vertebræ are the same in number, as a general rule, in all the species, being nineteen. The recognition of this important character appears to have been impeded by the variable number of moveable ribs in different species of the Artiodactyles, the dorsal vertebræ, which those ribs characterize, being fifteen in the Hippopotamus and twelve in the

Camel. -And the value of this distinction has been exaggerated owing to the common conception of the ribs as special bones distinct from the vertebræ, and their non-recognition as parts of a vertebra equivalent to the neurapophyses and other autogenous elements.

The vertebral formulæ of the Artiodactyle skeletons shew that the difference in the number of the so-called dorsal and lumbar vertebræ does not affect the number of the entire dorso-lumbar series: thus the Indian Wild Boar has $d. 13$, $l. 6 = 19$; the Domestic Hog and the Peccari have $d. 14, l. 5 = 19$; the Hippopotamus has $d. 15, l. 4 = 19$; the Gnu and Aurochs have $d. 14, l. 5 = 19$; the Ox and most of the true Ruminants have $d. 13, l. 6 = 19$; the aberrant Ruminants have $d. 12, l. 7 = 19$. The natural character and true affinities of the Artiodactyle group are further illustrated by the absence of the third trochanter in the femur, and by the place of perforation of the medullary artery at the fore and upper part of the shaft, as in the Hippopotamus, the Hog, and most of the Ruminants. The fore part of the astragalus is divided into two equal or subequal facets: the os magnum does not exceed, or is less than, the unciforme in size, in the carpus; and the ectocuneiform is less, or not larger, than the cuboid, in the tarsus. The digit answering to the third in the pentadactyle foot is unsymmetrical, and forms, with that answering to the fourth, a symmetrical pair. If the species be horned, the horns form one pair, as in most Ruminants, or two pairs, as in the four-horned Antelope and Sivathere; they are never developed singly, of symmetrical form, from the median line. The post-tympanic does not project downward distinctly from the mastoid, nor supersede it in any Artiodactyle; and the paroccipital always exceeds both those processes in length. The bony palate extends further back than in the Perissodactyles; the hinder aperture of the nasal passages is more vertical and commences posterior to the last molar tooth. The base of the pterygoid process is not perforated by the ectocarotid artery. The crowns of the premolars are smaller and less complex than those of the true molars, usually representing half of such crown. The last milk-molar is trilobed.

To these osteological and dental characters may be added some important modifications of internal structure, as, e. g., the complex form of the stomach in the Hippopotamus, Peccari, and Ruminants; the comparatively small and simple cæcum and the spirally folded colon in all Artiodactyles, which equally indicate the mutual affinities of the even-toed hoofed quadrupeds, and their claims to be regarded as a natural group of the *Ungulata*. The placenta is diffused in the Camel-tribe and non-ruminants; is cotyledonal in the true Ruminants. Many extinct genera, e. g. *Chœropotamus, Anthracotherium, Hyopotamus, Entelodon, Dichodon, Merycopotamus, Xiphodon, Dichobune, Anoplotherium, Microtherium,* &c., have been discovered, which once linked together the now broken series of Artiodactyles, represented by the existing genera, *Hippopotamus, Sus, Dicotyles, Camelus, Auchenia, Moschus, Camelopardalis, Cervus, Antilope, Ovis,* and *Bos.*

A well-marked, and at the present day very extensive subordinate group of the Artiodactyles, is called *Ruminantia*, in reference to the second mastication to which the food is subject after having been swallowed; the act of rumination requiring a peculiarly complicated form of stomach. The Ruminants have the 'cloven foot,' i. e. two hoofed digits on each foot forming a symmetrical pair, as by the cleavage of a single hoof; in most species there is added a pair of small supplementary hoofed toes. The metacarpals of the two functional toes coalesce to form a single 'cannon-bone,' as do the corresponding metatarsals. The Camel-tribe have the upper incisors reduced to a single pair; in the rest of the Ruminants the upper incisors are replaced by a callous pad. The lower canines are contiguous to the six lower incisors, and, save in the Camel-tribe, are similar to them, forming part of the same terminal series of eight teeth, between which and the molar series there is a wide interval. The true molars have their grinding surface marked by two double crescents, the convexity of which is turned inwards in the upper and outwards in the under jaw.

Many fossil Artiodactyles, with similar molars, appear to have differed from the Ruminants chiefly by retaining structures which are transitory and embryonic in most existing Ruminants, as, e. g. upper incisors and canines, first premolars, and separate metacarpal and metatarsal bones; these are among the lost links that once connected more intimately the Ruminants with the Hog and Hippopotamus.

The Pachyderms in the Cuvierian system included all the non-ruminant hoofed beasts; they were divided by the great French anatomist into the *Proboscidia, Solidungula,* and *Pachydermata ordinaria,* the latter again being subdivided according to the odd or even number of the hoofs. I have on another occasion[1] adduced evidence to shew that the right progression of the affinities of the *Ungulata* was broken by the interposition of the Horse and other Perissodactyles between the non-ruminant or omnivorous and the ruminant Artiodactyles; and that too high a value had been assigned to the Ruminantia by making them equivalent to all the other Ungulates collectively.

It is interesting, in relation to the needs of mankind, to find that, whilst some groups of *Ungulata,* e. g. the Perissodactyles and omnivorous Artiodactyles, have been gradually dying out, other groups, e. g. the Ruminants, have been augmenting in genera and species. Most interesting also is it to observe, that in existing Ungulates there is a more specialized structure, a further departure from the general type, than in their representatives of the miocene and eocene tertiary periods: such later and less typical Mammalia do more effective work by virtue of their adaptively modified structures.

The Ruminants, e. g., more effectually digest and assimilate grass, and form out of it a more nutritive and sapid kind of meat, than did the antecedent more typical and less specialized non-ruminant Herbivora.

[1] *Proceedings of the Geological Society,* November 3, 1847, p. 135.

The monodactyle Horse is a better and swifter beast of draught and burthen than its tridactyle predecessor the miocene *Hipparion* could have been. The nearer to a Tapir or a Rhinoceros in structure, the further would an equine quadruped be left from the goal in contending with a modern Racer.

With respect to the geographical distribution of the hoofed Mammalia, I may first remark that the order *Ruminantia* is principally represented by Old World species, of which 162 have been defined; only 24 species have been discovered in the New World, and none in Australia, New Guinea, New Zealand, or the Polynesian Isles.

The Camelopard is now peculiar to Africa; the Musk-deer to Africa and Asia: out of about 50 defined species of Antelope, only one is known in America, and none in the central and southern divisions of the New World. The Bison of North America is distinct from the Bison of Europe. The Musk-ox, peculiar for its limitation to high northern latitudes, is the sole bovine species that roams over the arctic coasts of both Asia and America. The Deer-tribe are more widely distributed. The Camels and Dromedaries of the Old World are represented by the Llamas and Vicugnas of the New. As, in regard to a former (tertiary) zoological period, the fossil *Camelidæ* of Asia are of the genus *Camelus*, so those of America are of the genus *Auchenia*. This geographical restriction ruled prior to any evidence of man's existence.

Palæontology has expanded our knowledge of the range of the Giraffe; during miocene or old pliocene periods, species of *Camelopardalis* roamed in Asia and Europe. Passing to the non-ruminant Artiodactyles, geology has also taught us that the Hippopotamus was not always confined, as now, to African rivers, but bathed, during pliocene times, in those of Asia and Europe. But no evidence has yet been had that the Giraffe or Hippopotamus were ever other than Old-World forms of *Ungulata*.

With respect to the Hog-tribe, we find that the true Swine (*Sus*) of the Old World are represented by Peccaries (*Dico-*

tyles) in the New; and geology has recently shewn that tertiary species of *Dicotyles* existed in North as well as South America. But no true *Sus* has been found fossil in either division of the New World, nor has any *Dicotyles* been found fossil in the Old World of the geographer. *Phacochœrus* (Wart-hogs) is a genus of the Hog-tribe at present peculiar to Africa.

The Rhinoceros is a genus now represented only in Asia and Africa; the species being distinct in the two continents. The islands of Java and of Sumatra have each their peculiar species; that of the latter being two-horned, as all the African Rhinoceroses are. Three or more species of two-horned Rhinoceros formerly inhabited Europe[1], one of which we know to have been warmly clad and adapted for a cold climate; but no fossil remains of the genus have been met with save in the Old World of the geographer. One of the earliest forms of European Rhinoceros was devoid of the nasal weapon: it has long been extinct.

Geology has given a wider prospect of the range of the Horse and Elephant, than was open to the student of living species only. The existing *Equidæ* and *Elephantidæ* properly belong, or are limited to, the Old World; and the Elephants to Asia and Africa, the species of the two continents being quite distinct. The horse, as Buffon remarked, carried terror to the eye of the indigenous Americans, viewing the animal for the first time, as it proudly bore their Spanish conqueror. But species of *Equus*, like species of *Mastodon*, coexisted with the *Megatherium* and *Megalonyx* in both South and North America, and perished with them, apparently before the human period.

The third division of the GYRENCEPHALA enjoy a higher degree of the sense of touch than the Ungulates through the greater number and mobility of the digits and the smaller extent to which they are covered by horny matter. This substance forms a single plate, in the shape of a claw or nail,

[1] See my *History of British Fossil Mammals*, 8vo, p. 350.

which is applied to only one of the surfaces of the extremity of the digit, leaving the other, usually the lower, surface possessed of its tactile faculty; whence the name *UNGUICULATA*, applied to this group, which, however, is here more restricted and natural than the group to which Linnæus extended the term. All the species are 'diphyodont,' and the teeth have a simple investment of enamel.

The first order, CARNIVORA, includes the beasts of prey, properly so called. With the exception of a few Seals the incisors are $\frac{3-3}{3-3}$ in number; the canines $\frac{1-1}{1-1}$, always longer than the other teeth, and usually exhibiting a full and perfect development as lethal weapons; the molars graduate from a trenchant to a tuberculate form, in proportion as the diet deviates from one strictly of flesh, to one of a more miscellaneous kind. The clavicle is rudimental or absent; the innermost digit is often rudimental or absent; they have no vesiculæ seminales; the teats are abdominal; the placenta is zonular.

The Carnivora are divided, according to modifications of the limbs, into 'pinnigrade,' 'plantigrade,' and 'digitigrade' tribes. In the Pinnigrades (Walrus, Seal-tribe) both fore and hind feet are short, and expanded into broad, webbed paddles for swimming, the hinder ones being fettered by continuation of integument to the tail. In the Plantigrades (Bear-tribe) the whole or nearly the whole of the hind foot forms a sole, and rests on the ground. In the Digitigrades (Cat-tribe, Dog-tribe, &c.) only the toes touch the ground, the heel being much raised.

It has been usual to place the Plantigrades at the head of the Carnivora, apparently because the higher order, Quadrumana, can put the heel to the ground: but the affinities of the Bear, as evidenced by internal structure, e. g. the renal and genital organs, are closer to the Seal-tribe; the broader and flatter pentadactyle foot of the plantigrade is nearer in form to the flipper of the seal than is the digitigrade, retractile-clawed, long and narrow hind foot of the feline quadruped, which is the highest and most typical of the Carnivora.

With the exception of the Dingo no true Carnivore exists in Australia, and that wild dog may have as little claim to be considered an autochthon as the low variety of Man, with whom it is sometimes associated in a half-tamed state.

The genus *Ursus* is represented by species indigenous to Europe, Asia, Africa, and America; but those of the temperate and warmer latitudes of the New World are distinct from the species of the Old World. Certain plantigrade genera, e. g. *Procyon* (Racoons), *Nasua* (Coati-mondis) and *Cercoleptes* (Kinkajous) are peculiarly American: other plantigrade genera, e. g. *Mydaus*, *Ailurus*, and *Arctictis*, are peculiarly Asian.

The genus Hyæna is limited to the Old World, and one species (*H. crocuta*) to Southern Africa.

The Skunks (*Mephitis*) are peculiar to America; the viverrine Carnivores to the Old World.

The great fulvous felines (*Leo*) of Africa and Asia are represented in America by the smaller Pumas: the Old World spotted felines by the Jaguars: the great striped felines (*Tigris*) are now restricted to Asia.

The principle of the more specialized character of actual organisations receives illustration in the genetic history of the present order.

The genera *Felis* and *Machairodus*, with their curtailed and otherwise modified dentition and their strong short jaws, become, thereby, more powerfully and effectively destructive than the eocene *Hyænodons* and miocene *Pterodons*, with their numerically typical dentition and their three carnassial teeth on each side of the concomitantly prolonged jaws, could have been.

In the most strictly carnivorous *GYRENCEPHALA* the paw is perfected as an instrument for retaining and lacerating a struggling prey by the superadded elastic structures for retracting the claws and maintaining them sharp. We next find in the unguiculate limb such a modification in the size, shape, position, and direction of the innermost digit that it can be opposed, as a thumb, to the other digits, thus constituting what is properly termed a 'hand.' Those Unguiculates which have both fore and hind limbs so modified, form the order QUAD-RUMANA. They have $\frac{2-2}{2-2}$ incisors, and $\frac{3-3}{3-3}$ broad tuberculate molars; perfect clavicles; pectoral mammæ; vesicular and prostatic glands; a discoid, sometimes double, placenta. The Quadrumana have a well-marked threefold geographical as well as structural division.

The Strepsirhines are those with curved or twisted terminal nostrils, with much modified incisors, commonly $\frac{3-3}{3-3}$; premolars $\frac{3-3}{3-3}$ or $\frac{2-2}{2-2}$ in number, and molars with sharp tubercles: the second digit of the hind limb has a claw. This group includes the Galagos, Pottos, Loris, Aye-Ayes, Indris, and the true Lemurs; the three latter genera being restricted to Madagascar, whence the group diverges in one direction to the continent of Africa, in the other to the Indian Archipelago.

The Platyrhines are those with the nostrils subterminal and wide apart; premolars $\frac{3-3}{3-3}$ in number, the molars with blunt tubercles; the thumbs of the fore-hands not opposable or wanting; the tail in most prehensile; they are peculiar to South America.

The Catarhines have the nostrils oblique and approximated below, and opening above and behind the muzzle: the premolars are $\frac{2-2}{2-2}$ in number; the thumb of the fore-hand is opposable. They are restricted to the Old World, and, save a single species on the rock of Gibraltar, to Africa and Asia. The highest organized family of Catarhines is tailless, and offers in the Orang, Chimpanzee, and Gorilla, the nearest approach to the human type.

The Catarhine monkeys include the Macaques, most of which are Asiatic, a few are African, and one European; the Cercopitheques, most of which are African, and a few Asiatic; and other genera which characterize one or other continent exclusively. Thus the true Baboons (*Papio*) are African, as are the thumbless Monkeys (*Colobus*) and the Chimpanzees (*Troglodytes*). The Semnopitheques, Gibbons (*Hylobates*), and Orangs (*Pithecus*) are peculiarly Asiatic. Palæontology has shown that a Macaque, a Gibbon and an Orang existed during the older tertiary times in Europe; and that a Semnopitheque existed in miocene times in India. But all the fossil remains of Quadrumana in the Old World belong to the family *Catarhina*, which is still exclusively confined to that great division of dry land. The tailless Macaque (*Inuus sylvanus*) of Gibraltar may have existed in that part of the Old World before Europe was separated by the Straits of Gibraltar from Africa. Fossil remains of Quadrumana have been discovered in South America; they indicate Platyrhine forms: a species, for example, allied to the Howlers (*Mycetes*), but larger than any now known to exist, has left its remains in Brazil.

Whilst adverting to the geographical distribution of Quadrumana, I would contrast the peculiarly limited range of the Orangs and Chimpanzees with the cosmopolitan powers of mankind. The two species of Orang (*Pithecus*) are confined to Borneo and Sumatra; the two species of Chimpanzee (*Troglodytes*) are limited to an intertropical tract of the western part of Africa. They appear to be inexorably bound to their localities by climatal influences regulating the assemblage of certain trees and the production of certain fruits. With all our care, in regard to choice of food, clothing, and contrivances for artificially maintaining the chief physical conditions of their existence, the healthiest specimens of Orang or Chimpanzee, brought over in the vigour of youth, perish within a period never exceeding three years, and usually much shorter, in our climate. By what metamorphoses, we may ask, has the alleged humanized Chimpanzee or Orang been brought to endure all climates? The advocates of 'transmutation' have failed to explain them. Certain it is that those physical differences in cerebral, dental, and osteo-logical structure, which place, in my estimate of them, the genus *Homo* in a distinct group of the Mammalian class, zoo-logically of higher value than the 'order,' are associated with equally contrasted powers of endurance of different climates, whereby Man has become a denizen of every part of the globe from the torrid to the arctic zones.

Climate rigidly limits the range of the Quadrumana in latitude: creational and geographical causes limit their range in longitude. Distinct genera represent each other in the same latitudes of the New and Old Worlds; and also, in a great degree, in Africa and Asia. But the development of an Orang out of a Chimpanzee, or reciprocally, is physiologically incon-ceivable. (Appendix B).

The sole representative of the ARCHENCEPHALA, is the ge-nus *Homo*. His structural modifications, more especially of the lower limb, by which the erect stature and bipedal gait are maintained, are such as to claim for Man ordinal distinc-tion on merely external zoological characters. But, as I have already argued, his mental powers, in association with his extraordinarily developed brain, entitle the group which he represents to equivalent rank with the other primary divi-sions of the class *Mammalia* founded on cerebral characters. In this primary group Man forms but one genus, *Homo*, and that genus but one order, called BIMANA, on account of the opposable thumb being restricted to the upper pair of limbs. The mammæ are pectoral. The placenta is a single, sub-circular, cellulo-vascular, discoid body.

Man has only a partial covering of hair, which is not merely protective of the head, but is ornamental and distinc-tive of sex. The dentition of the genus *Homo* is reduced to thirty-two teeth by the suppression of the outer incisor and

E

the first two premolars of the typical series on each side of both jaws, the dental formula being :—

$$i. \frac{2-2}{2-2}, \quad c. \frac{1-1}{1-1}, \quad p. \frac{2-2}{2-2}, \quad m. \frac{3-3}{3-3} = 32.^{1}$$

All the teeth are of equal length, and there is no break in the series ; they are subservient in Man not only to alimentation, but to beauty and to speech.

The human foot is broad, plantigrade, with the sole, not inverted as in Quadrumana, but applied flat to the ground; the leg bears vertically on the foot ; the heel is expanded beneath ; the toes are short, but with the innermost longer and much larger than the rest, forming a ' hallux' or great toe, which is placed on the same line with, and cannot be opposed to, the other toes ; the pelvis is short, broad, and wide, keeping the thighs well apart; and the neck of the femur is long, and forms an open angle with the shaft, increasing the basis of support for the trunk. The whole vertebral column, with its slight alternate curves, and the well-poised, short, but capacious subglobular skull, are in like harmony with the requirements of the erect position. The widely-separated shoulders, with broad scapulæ and complete clavicles, give a favourable position to the upper limbs, now liberated from the service of locomotion, with complex joints for rotatory as well as flexile movements, and terminated by a hand of matchless perfection of structure, the fit instrument for executing the behests of a rational intelligence and a free will. Hereby, though naked, Man can clothe himself, and rival all natural vestments in warmth and beauty ; though defenceless, Man can arm himself with every variety of weapon, and become the most terribly destructive of animals. Thus he fulfils his destiny as the master of this earth, and of the lower Creation.

Such are the dominating powers with which we, and we alone, are gifted ! I say gifted, for the surpassing organisation was no work of ours. It is He that hath made us ; not

[1] Vid. p. 19, for the type-formula and explanation of its symbols.

we ourselves. This frame is a temporary trust, for the uses of which we are responsible to the Maker.

Oh! you who possess it in all the supple vigour of lusty youth, think well what it is that He has committed to your keeping. Waste not its energies; dull them not by sloth: spoil them not by pleasures! The supreme work of Creation has been accomplished that you might possess a body—the sole erect—of all animal bodies the most free—and for what? for the service of the soul.

Strive to realise the conditions of the possession of this wondrous structure. Think what it may become—the Temple of the Holy Spirit! Defile it not. Seek, rather, to adorn it with all meet and becoming gifts, with that fair furniture, moral and intellectual, which it is your inestimable privilege to acquire through the teachings and examples and ministrations of this Seat of Sound Learning and Religious Education.

Such, Sir, are the sentiments that have naturally flowed from the contemplation of the highest of the gradations of Mammalian structure of which we have now completed the ascensive survey: and I know of no topic more fitting to the words in which, with a grateful sense of the most kind reception and attention accorded to me, I conclude the duty of this Chair.

TABLE OF THE SUBCLASSES AND ORDERS OF THE MAMMALIA, ACCORDING TO THE CEREBRAL SYSTEM.

CLASS.	SUBCLASS.		ORDER.	GENUS OR FAMILY.	EXAMPLE.
MAMMALIA	Archencephala		BIMANA	Homo	Man.
	Gyrencephala	Unguiculata	QUADRUMANA	Catarhina	Ape.
				Platyrhina	Marmoset.
				Strepsirhina	Lemur.
			CARNIVORA	Digitigrada	Dog.
				Plantigrada	Bear.
				Pinnigrada	Seal.
		Ungulata	ARTIODACTYLA	Omnivora	Hog.
				Ruminantia	Sheep.
				Solidungula	Horse.
			PERISSODACTYLA	Multungula	Tapir.
			PROBOSCIDIA	Elephas	Elephant.
				Dinotherium.	
			TOXODONTIA	Toxodon.	
				Nesodon.	
		Mutilata	SIRENIA	Manatus	Sea-cow.
				Halicore	Dugong.
			CETACEA	Delphinidæ	Porpoise.
				Balenidæ	Whale.
	Lissencephala		BRUTA	Bradypodidæ	Sloth.
				Dasypodidæ	Armadillo.
				Edentula	Anteater.
			CHEIROPTERA	Frugivora	Roussette.
				Insectivora	Bat.
			INSECTIVORA	Talpidæ	Mole.
				Erinaceidæ	Hedgehog.
				Soricidæ	Shrew.
			RODENTIA	Non-claviculata	Hare.
				Claviculata	Rat.
	Lyencephala		MARSUPIALIA	Rhizophaga	Wombat.
				Poëphaga	Kangaroo.
				Carpophaga	Phalanger.
				Entomophaga	Opossum.
			MONOTREMATA	Echidna	Echidna.
				Ornithorhynchus	Duck-mole.

APPENDIX.

APPENDIX.

APPENDIX A.

ON THE EXTINCTION OF SPECIES.

*Being the Conclusion of the Fullerian Course of Lectures on
Physiology, for* 1859.

IN a *Report* to the British Association for the Advancement of
Science, *On the Extinct Mammals of Australia*, published in the
Volume of Reports for 1844, evidence is adduced in proof of the
law, that with extinct as with existing mammalia particular forms
were assigned to particular provinces, and that the same forms
were restricted to the same provinces at a former geological period
as they are at the present day. That period, however, was the
more recent tertiary one.

In carrying back the retrospective comparison of existing and
extinct mammals to those of the eocene and oolitic strata, in relation to their local distribution, we obtain indications of extensive
changes in the relative position of sea and land during those epochs,
through the degree of incongruity between the generic forms of the
mammalia which then existed in Europe, and any that actually
exist on the great natural continent of which Europe now forms
part. It would seem, indeed, that the further we penetrate into
time for the recovery of extinct mammalia, the further we must
go into space to find their existing analogues. To match the eocene palæotheres and lophiodons we must bring tapirs from Sumatra or South America; and we must travel to the antipodes for
myrmecobians, the nearest living analogue to the amphitheres and
spalacotheres of our oolitic strata.

On the problem of the extinction of species I have little to say;
and of the more mysterious subject of their coming into being,
nothing profitable or to the purpose. As a cause of extinction in
times anterior to man, it is most reasonable to assign the chief
weight to those gradual changes in the conditions affecting a due
supply of sustenance to animals in a state of nature which must
have accompanied the slow alternations of land and sea brought

about in the æons of geological time. Yet this reasoning is applicable only to land-animals; for it is scarcely conceivable that such operations can have affected sea-fishes.

There are characters in land-animals rendering them more obnoxious to extirpating influences, which may explain why so many of the larger species of particular groups have become extinct, whilst smaller species of equal antiquity have survived. In proportion to its bulk is the difficulty of the contest which the animal has to maintain against the surrounding agencies that are ever tending to dissolve the vital bond, and subjugate the living matter to the ordinary chemical and physical forces. Any changes, therefore, in such external agencies as a species may have been originally adapted to exist in, will militate against that existence in a degree proportionate to the size which may characterise the species. If a dry season be gradually prolonged, the large mammal will suffer from the drought sooner than the small one; if such alteration of climate affect the quantity of vegetable food, the bulky herbivore will first feel the effects of stinted nourishment; if new enemies be introduced, the large and conspicuous animal will fall a prey while the smaller kinds conceal themselves and escape. Small quadrupeds, moreover, are more prolific than large ones. Those of the bulk of the mastodons, megatheria, glyptodons, and diprotodons, are uniparous. The actual presence, therefore, of small species of animals in countries where larger species of the same natural families formerly existed, is not the consequence of degeneration—of any gradual diminution of the size—of such species, but is the result of circumstances which may be illustrated by the fable of the 'Oak and the Reed;' the smaller and feebler animals have bent and accommodated themselves to changes to which the larger species have succumbed.

That species should become extinct appears, from the abundant evidence of the fact of extinction, to be a law of their existence; whether, however, it be inherent in their own nature, or be relative and dependent on inevitable changes in the conditions and theatre of their existence, is the main subject for consideration. But, admitting extinction as a natural law which has operated from the beginning of life on this planet, it might be expected that some evidence of it should occur in our own time, or within the historical period. Reference has been made to several instances of the extirpation of species, certainly, probably, or possibly, due to the direct agency of man; but this cause avails not in the question of the extinction of species at periods prior to any evidence of human existence; it does not help us in the explanation of the majority of extinctions; as of the races of aquatic invertebrata which have successively passed away.

Within the last century academicians of St. Petersburg and good naturalists have described and given figures of the bony and the perishable parts, including the alimentary canal, of a large and peculiar fucivorous Sirenian—an amphibious animal like the Manatee, which Cuvier classified with his herbivorous *Cetacea*, and called *Stelleria*, after its discoverer. This animal inhabited the Siberian shores and the mouths of the great rivers there disemboguing. It is now believed to be extinct, and this extinction seems not to have been due to any special quest and persecution by man. We may discern, in this fact, the operation of changes in physical geography which have, at length, so affected the conditions of existence of the *Stelleria* as to have caused its extinction. Such changes had operated, at an earlier period, to the extinction of the Siberian elephant and rhinoceros of the same regions and latitudes. A future generation of zoologists may have to record the final disappearance of the Arctic buffalo (*Ovibos moschatus*). Fossil remains of *Ovibos* and *Stelleria* shew that they were contemporaries of *Elephas primigenius* and *Rhinoceros tichorrhinus*.

The Great Auk (*Alca impennis*, L.) seems to be rapidly verging to extinction. It has not been specially hunted down, like the dodo and dinornis, but by degrees has become more scarce. Some of the geological changes affecting circumstances favourable to the well-being of the *Alca impennis*, have been matters of observation. A friend[1], who last year visited Iceland, informs me that the last great auks, known with anything like certainty to have been there seen, were two which were taken in 1844 during a visit made to the high rock called 'Eldey,' or 'Meelsoekten,' lying off Cape Reykianes, the S. W. point of Iceland. This is one of three principal rocky islets formerly existing in that direction, of which the one, specially named from this rare bird 'Geirfugla Sker,' sank to the level of the surface of the sea during a volcanic disturbance in or about the year 1830. Such disappearance of the fit and favourable breeding-places of the *Alca impennis* must form an important element in its decline towards extinction. The numbers of the bones of *Alca impennis* on the shores of Iceland, Greenland, and Denmark, attest the abundance of the bird in

[1] John Wolley, jun., Esq. F.Z.S.

former times. A consideration of such instances of modern partial or total extinctions may best throw light on, and suggest the truest notions of, the causes of ancient extinctions.

As to the successions, or coming in, of new species, one might speculate on the gradual modifiability of the individual; on the tendency of certain varieties to survive local changes, and thus progressively diverge from an older type; on the production and fertility of monstrous offspring; on the possibility, *e. g.* of a variety of auk being occasionally hatched with a somewhat longer winglet, and a dwarfed stature; on the probability of such a variety better adapting itself to the changing climate or other conditions than the old type—of such an origin of *Alca torda, e. g.*;—but to what purpose? Past experience of the chance aims of human fancy, unchecked and unguided by observed facts, shews how widely they have ever glanced away from the gold centre of truth.

The sum of the evidence which has been obtained appears to prove that the successive extinction of *Amphitheria, Spalacotheria, Triconodons*, and other mesozoic forms of mammals, has been followed by the introduction of much more numerous, varied, and higher-organised forms of the class, during the tertiary periods.

There are, however, geologists who maintain that this is an assumption, based upon a partial knowledge of the facts. Mere negative evidence, they allege, can never satisfactorily establish the proposition that the mammalian class is of late introduction, nor prevent the conjecture that it may have been as richly represented in secondary as in tertiary times, could we but get evidence of the terrestrial fauna of the oolitic continent. To this objection I have to reply : in the palæozoic strata, which, from their extent and depth, indicate, in the earth's existence as a seat of organic life, a period as prolonged as that which has followed their deposition, no trace of mammals has been observed. It may be conceded that, were mammals peculiar to dry land, such negative evidence would weigh little in producing conviction of their non-existence during the Silurian and Devonian æons, because the explored parts of such strata have been deposited from an ocean, and the chance of finding a terrestrial and air-breathing creature's remains in oceanic deposits is very remote. But, in the present state of the warm-blooded, air-breathing, viviparous class, no genera and species are represented by such numerous and widely dispersed individuals, as those of the order *Cetacea*, which, under the guise of fishes, dwell, and can only live, in the ocean.

In all cetacea the skeleton is well ossified, and the vertebræ are very numerous : the smallest cetaceans would be deemed large amongst land-mammals; the largest surpass in bulk any creatures of which we have yet gained cognizance : the hugest ichthyosaur, iguanodon, megalosaur, mammoth, or megathere, is a dwarf in comparison with the modern whale of a hundred feet in length.

During the period in which we have proof that *Cetacea* have existed, the evidence in the shape of bones and teeth, which latter enduring characteristics in most of the species are peculiar for their great number in the same individual, must have been abundantly deposited at the bottom of the sea; and as cachalots, grampuses, dolphins, and porpoises are seen gambolling in shoals in deep oceans, far from land, their remains will form the most characteristic evidences of vertebrate life in the strata now in course of formation at the bottom of such oceans. Accordingly, it consists with the known characteristics of the cetacean class to find the marine deposits which fell from seas tenanted, as now, with vertebrates of that high grade, containing the fossil evidences of the order in vast abundance.

The red crag of our eastern counties contains petrified fragments of the skeletons and teeth of various *Cetacea*, in such quantities as to constitute a great part of that source of phosphate of lime for which the red crag is worked for the manufacture of artificial manure. The scanty and dubious evidence of *Cetacea* in newer secondary beds[1] seems to indicate a similar period for their beginning as for the soft-scaled cycloid and ctenoid fishes which have superseded the ganoid orders of mesozoic times.

We cannot doubt but that had the genera *Ichthyosaurus*, *Pliosaurus*, or *Plesiosaurus*, been represented by species in the same ocean that was tempested by the Balænodons and Dioplodons of the miocene age, the bones and teeth of those marine reptiles would have testified to their existence as abundantly as they do at a previous epoch in the earth's history. But no fossil relic of an enaliosaur has been found in tertiary strata, and no living enaliosaur has been detected in the present seas : and they are consequently held by competent naturalists to be extinct.

In like manner does such negative evidence weigh with me in proof of the non-existence of marine mammals in the liassic and oolitic times. In the marine deposits of those secondary or meso-

[1] See 'Introduction' to Owen's *History of British Fossil Mammalia*, 8vo., 1846, p. xv.

zoic epochs, the evidence of vertebrates governing the ocean, and preying on inferior marine vertebrates, is as abundant as that of air-breathing vertebrates in the tertiary strata; but in the one the fossils are exclusively of the cold-blooded reptilian class, in the other, of the warm-blooded mammalian class. The *Enaliosauria*, *Cetiosauria*, and *Crocodilia*, played the same part and fulfilled similar offices in the seas from which the lias and oolites were precipitated, as the *Delphinidæ* and *Balænidæ* did in the tertiary, and still do in the present, seas. The unbiassed conclusion from both negative and positive evidence in this matter is, that the *Cetacea* succeeded and superseded the *Enaliosauria*. To the mind that will not accept such conclusion, the stratified oolitic rocks must cease to be monuments or trustworthy records of the condition of life on the earth at that period.

So far, however, as any general conclusion can be deduced from the large sum of evidence above referred to, and contrasted, it is against the doctrine of the Uniformitarian. Organic remains, traced from their earliest known graves, are succeeded, one series by another, to the present period, and never re-appear when once lost sight of in the ascending search. As well might we expect a living Ichthyosaur in the Pacific, as a fossil whale in the Lias : the rule governs as strongly in the retrospect as the prospect. And not only as respects the *Vertebrata*, but the sum of the animal species at each successive geological period has been distinct and peculiar to such period.

Not that the extinction of such forms or species was sudden or simultaneous : the evidences so interpreted have been but local : over the wider field of life at any given epoch, the change has been gradual; and, as it would seem, obedient to some general, but as yet, ill-comprehended law. In regard to animal life, and its assigned work on this planet, there has, however, plainly been ‘an ascent and progress in the main.’

Although the mammalia, in regard to the plenary development of the characteristic orders, belong to the Tertiary division of geological time, just as ‘ *Echini* are most common in the superior strata, *Ammonites* in those beneath, and *Producti* with numerous *Encrini* in the lowest’[1] of the secondary strata, yet the beginnings of the class manifest themselves in the formations of the earlier preceding division of geological time.

No one, save a prepossessed Uniformitarian, would infer from

[1] A generalisation of WILLIAM SMITH'S.

the *Lucina* of the permian, and the *Opis* of the trias, that the Lamellibranchiate Mollusks existed in the same rich variety of development at these periods as during the tertiary and present times; and no prepossession can close the eyes to the fact that the Lamellibranchiate have superseded the Palliobranchiate bivalves.

On negative evidence *Orthisina, Theca, Producta,* or *Spirifer* are believed not to exist in the present seas : neither are the existing genera of siphonated bivalves and univalves deemed to have abounded in permian, triassic or oolitic times. To suspect that they may have then existed, but have hitherto escaped observation, because certain Lamellibranchs with an open mantle, and some holostomatous and asiphonate Gastropods, have left their remains in secondary strata, is not more reasonable, as it seems to me, than to conclude that the proportion of mammalian life may have been as great in secondary as in tertiary strata, because a few small forms of the lowest orders have made their appearance in triassic and oolitic beds.

Turning from a retrospect into past time for the prospect of time to come,—and I have received more than one inquiry into the amount of prophetic insight imparted by Palæontology—I may crave indulgence for a few words, of more sound, perhaps, than significance. But the reflective mind cannot evade or resist the tendency to speculate on the future course and ultimate fate of vital phenomena in this planet.

There seems to have been a time when life was not; there may, therefore, be a period when it will cease to be.

Our most soaring speculations still shew a kinship to our nature : we see the element of finality in so much that we have cognizance of, that it must needs mingle with our thoughts, and bias our conclusions on many things.

The end of the world has been presented to man's mind under divers aspects:—as a general conflagration ; as the same, preceded by a millennial exaltation of the world to a Paradisiacal state,— the abode of a higher and blessed race of intelligences.

If the guide-post of Palæontology may seem to point to a course ascending to the condition of the latter speculation, it points but a very short way, and in leaving it we find ourselves in a wilderness of conjecture, where to try to advance is to find ourselves 'in wandering mazes lost.'

With much more satisfaction do I return to the legitimate deductions from the phenomena we have had under review.

In the survey which I have taken in the present course of lectures of the genesis, succession, geographical distribution, affinities, and osteology of the mammalian class, if I have succeeded in demonstrating the perfect adaptation of each varying form to the exigencies, and habits, and well-being of the species, I have fulfilled one object which I had in view, viz. to set forth the beneficence and intelligence of the Creative Power.

If I have been able to demonstrate a uniform plan pervading the osteological structure of so many diversified animated beings, I must have enforced, were that necessary, as strong a conviction of the unity of the Creative Cause.

If, in all the striking changes of form and proportion which have passed under review, we could discern only the results of minor modifications of the same few osseous elements,—surely we must be the more strikingly impressed with the wisdom and power of that Cause which could produce so much variety, and at the same time such perfect adaptations and endowments, out of means so simple.

For, in what have those mechanical instruments,—the hands of the ape, the hoofs of the horse, the fins of the whale, the trowels of the mole, the wings of the bat,—so variously formed to obey the behests of volition in denizens of different elements—in what, I say, have they differed from the artificial instruments which we ourselves plan with foresight and calculation for analogous uses, save in their greater complexity, in their perfection, and in the unity and simplicity of the elements which are modified to constitute these several locomotive organs?

Everywhere in organic nature we see the means not only subservient to an end, but that end accomplished by the simplest means. Hence we are compelled to regard the Great Cause of all, not like certain philosophic ancients, as a uniform and quiescent mind, as an all pervading *anima mundi*, but as an active and anticipating intelligence.

By applying the laws of comparative anatomy to the relics of extinct races of animals contained in and characterizing the different strata of the earth's crust, and corresponding with as many epochs in the earth's history, we make an important step in advance of all preceding philosophies, and are able to demonstrate that the same pervading, active, and beneficent intelligence which manifests His power in our times, has also manifested His power in times long anterior to the records of our existence.

But we likewise, by these investigations, gain a still more important truth, viz. that the phenomena of the world do not succeed each other with the mechanical sameness attributed to them in the cycles of the Epicurean philosophy; for we are able to demonstrate that the different epochs of the earth were attended with corresponding changes of organic structure; and that, in all these instances of change, the organs, as far as we could comprehend their use, were exactly those best suited to the functions of the being. Hence we not only show intelligence evoking means adapted to the end; but, at successive times and periods, producing a change of mechanism adapted to a change in external conditions. Thus the highest generalizations in the science of organic bodies, like the Newtonian laws of universal matter, lead to the unequivocal conviction of a great First Cause, which is certainly not mechanical.

Unfettered by narrow restrictions,—unchecked by the timid and unworthy fears of mistrustful minds, clinging, in regard to mere physical questions, to beliefs, for which the Author of all truth has been pleased to substitute knowledge,—our science becomes connected with the loftiest of moral speculations.

If I believed,—to use the language of a gifted contemporary,—that the imagination, the feelings, the active intellectual powers, bearing on the business of life, and the highest capacities of our nature, were blunted and impaired by the study of physiological and palæontological phenomena, I should then regard our science as little better than a moral sepulchre, in which, like the strong man, we were burying ourselves and those around us in ruins of our own creating.

But surely we must all believe too firmly in the immutable attributes of that Being, in whom all truth, of whatever kind, finds its proper resting-place, to think that the principles of physical and moral truth can ever be in lasting collision[1].

[1] Sedgwick, *Address to the Geological Society*, 1831.

APPENDIX B.

ON THE ORANG, CHIMPANZEE, AND GORILLA,

With reference to the 'Transmutation of Species.'

FOR about two centuries, naturalists have been cognizant of a small ape, tailless, without cheek-pouches, and without the ischial callosities, clothed with black hair, with a facial angle of about 60°, and of a physiognomy milder and more human-like than in,the ordinary race of monkeys, less capricious, less impulsive in its habits, more staid and docile. This species, brought from the West Coast of Africa, is that which our anatomist, Tyson, dissected: he described the main features of its organisation in his work published in 1699[1]. He called it the *Homo Sylvestris*, or pigmy. It is noted by Linnæus, in some editions of his *Systema Naturæ*, as the *Homo Troglodytes*. Blumenbach, giving a truer value to the condition of the innermost digit of the hind foot, which was like a thumb, called it the *Simia Troglodytes;* it afterwards became more commonly known as the 'Chimpanzee.'

At a later period, naturalists became acquainted with a similar kind of ape, of quiet docile disposition, with the same sad, human-like expression of features. It was brought from Borneo or Sumatra; where it is known by the name of *Orang*, which, in the language of the natives of Borneo, signifies 'man,' with the distinctive addition of *Outan*, meaning 'Wood-man,' or 'Wild Man of the Woods.' This creature differed from the pigmy, or *Simia Troglodytes* of Africa, by being covered with hair of a reddish-brown colour, and by having the anterior, or upper limbs, much longer in proportion, and the thumb upon the hind feet somewhat less. It was entered in the zoological catalogue as the *Simia Satyrus*. A governor of Batavia, Baron Wurmb, had transmitted to Holland, in 1780, the skeleton of a large kind of ape, tailless, like this small species from Borneo, but with a much-developed face, and large canine teeth, and bearing thick callosities upon the cheeks, giving it, upon the whole, a very baboon-like physiognomy; and he called it the *Pongo.*

At the time when Cuvier revised his summary of our knowledge of the animal kingdom, in the second edition of his *'Règne Animal,'*

[1] *'Orang-Outang, sive Homo sylvestris;* or the Anatomie of a Pygmie, compared with that of a Monkey, an Ape and a Man,' 4to, 1699.

1829, the knowledge of the anthropoid apes was limited to these three forms. It had been suspected that the pongo might be the adult form of the orang; but Cuvier, pointing to its distinctive characters, suggested that it could hardly be the same species. The facial angle of the small red orang of Borneo, and of the small black chimpanzee of Africa, brought them, from the predominant cranium, and small size of the jaws and small teeth, nearer than any other known mammalian animal to the human species, particularly to the lower, or negro forms. It was evident, from the examination of these small chimpanzees and orangs, that they were the young of some large species of ape. The small size and number of their teeth, (there being, in some of the smaller specimens, only twenty, like the number of deciduous teeth in the human species,) and the intervals between those teeth, all showed them to be of the first or deciduous series. In 1835 I availed myself of the rich materials in regard to these animals collected about that time by the Zoological Society, to investigate the state of dentition, especially that of the permanent teeth which might be hidden in the substance of the jaws, of both the immature orang-outang and the chimpanzee, and I found that the germs of those teeth in the orang-outang agreed in size with the permanent teeth that were developed in the jaws of a species of the pongo of Wurmb, which Sir Stamford Raffles had presented to the museum of the College of Surgeons some years before. Specimens of orangs since acquired, of an intermediate age, have shown the progressive change of the dentition.

In the substance of the jaw were found the germs of the great canines, and of large bicuspid teeth; foreshowing the changes that must take place when the jaw is sufficiently enlarged to receive permanent teeth of this kind; and, when the rest of the cranium is modified, concomitantly, for the attachment of muscles to work the jaw so armed, denoting that all these changes must result in the acquisition of characters such as are presented by the skulls of the large pongo, or Bornean baboon-like ape. The specific identity of the pongo with certain of the young orang-outangs, was thus satisfactorily made out, and is now admitted by all naturalists. With regard to the chimpanzee, the germs of similarly proportioned large teeth were also discovered in the jaws, indicating, in like manner, that it must be the young of a much larger species of ape.

The principal osteological characters of the chimpanzee and

F

orang, commencing from the vertebral column, are as follows :—
The vertebral column describes only one curve, inclining forward,
where it supports the head with its large jaws and teeth. The
vertebræ in the neck, seven in number as usual in the mammalia,
are chiefly remarkable for the great length of the simple spinous
processes developed more than in most of the inferior apes, in
relation to the necessities of the muscular masses that are to sustain and balance the head that preponderates so much forward on
the neck. The vertebræ maintain a much closer correspondence
in size, from the cervical to the dorsal and lumbar region, than in
the human skeleton. With regard to the dorsal vertebræ, or those
to which moveable ribs are articulated, there are twelve pairs in the
orang; seven of them send cartilages to join the sternum, which is
more like the sternum in man than in any of the inferior quadrumana: it is shorter and broader. In the smaller long-armed apes
(*Hylobates*), which make the first step in the transition from the
ordinary quadrumana to the man-like apes, the sternum is remarkably broad and short. The lumbar vertebræ are, originally, five
in number in the orang; but one or two may coalesce with the
sacrum. The sacrum is broader than in the lower quadrumana,
but it is still narrow in comparison with its proportions in man.
The pelvis is longer. The iliac bones are more expanded than
in the lower quadrumana, but on the same plane, and are flattened and long. The tuberosities of the ischia are remarkably
developed, and project outward. All these conditions of the vertebral column indicate an animal capable only of a semi-erect
position, and present a modification of the trunk much better
adapted for a creature destined for a life in trees, than one that is
to walk habitually erect upon the surface of the ground. But
that adaptation of the skeleton is still more strikingly shown in
the unusual development of the upper prehensile extremities. The
scapula is broad, with a well-developed spine and acromion; there
is a complete clavicle; the bone of the arm (*humerus*) is of remarkable length, in proportion to the trunk; the radius and the ulna
are also very long, and unusually diverging, to give increased surface of attachment to muscles; the hand is remarkable for the
length of the metacarpus, and of the phalanges, which are slightly
bent towards the palm; the thumb is less developed than the corresponding digit in the foot; the whole hand is admirably adapted
for retaining a firm grasp of the boughs of trees. In the structure
of the carpus, there is a well-marked difference from the human

subject, and a retention of the character met with in the lower quadrumana; the scaphoid bone being divided in the orang-outang. In the chimpanzee the bones of the carpus are eight, as in the human subject, but differ somewhat in form. If the upper extremities are so extraordinary for their disproportionate length, the lower ones are equally remarkable for their disproportionate smallness in comparison with the trunk, in the orang. The femur is short and straight, and the neck of the thigh-bone comparatively short. The head of the thigh-bone in this animal, which requires the use of these lower prehensible organs to grasp the branches of trees, and to move freely in many directions, is free from that ligament which strengthens the hip-joint in man; the head of the femur in the orang is quite smooth, without any indication of that attachment. Here, again, the chimpanzee manifests a nearer approach to man, for the *ligamentum teres* is present in it in accordance with the stronger and better development of the whole hind-limb. This approximation, also, is more especially marked in the larger development of the innermost of the five digits of the foot in the chimpanzee, which is associated with a tendency to move more frequently upon the ground, to maintain a more erect position than the orang-outang, and to walk further without the assistance of a stick. The foot, in both these species of anthropoid orangs, is characterized by the backward position of the ankle-joint surface presented by the astragalus to the tibia, which serves for the transference of the superincumbent weight upon the foot; by the comparatively feeble development of the backward projecting process of the calcaneum; by the obliquity of the articular surface of the astragalus, which tends to incline the foot a little inwards, taking away from the plantigrade character of the creatures and from their capacity to support themselves in an erect position, and giving them an equivalent power of applying their prehensile feet to the branches of the trees in which they live.

In both the orang and chimpanzee the skull is articulated to the spine by condyles, which are placed far back on its under surface. The cranium is small, characterised by well-developed occipital and sagittal ridges; the occipital ridges in reference to the muscles sustaining the head; and the sagittal ones in reference to an increased extent of the temporal muscles. The zygomatic arches are strong, and well arched outwards. The lower jaw is of great depth, and has powerful ascending rami, but the chin is wanting. The facial angle is about 50° to 55° in the full-grown *Simia satyrus*, and

55^0 to 60^0 in the *Troglodytes niger*. The difference in the facial angle between the young and adult apes, (which, in the young chimpanzee, approaches 60^0 to 65^0,) depends upon those changes consequent upon the shedding of the deciduous teeth and the concomitant development of the jaws and intermuscular processes of the cranium.

But the knowledge of the species of these anthropoid apes has been further increased since the acquisition of a distinct and precise cognisance of the characters of the adults of the orang and chimpanzee. First, in reference to the orangs of Borneo, specimens have reached this country which show that there is a smaller species in that island, the *Simia Morio*, in which the canines are less developed, in which the bony *cristæ* are never raised above the level of the ordinary convexity of the cranium, and in which the callosities upon the cheeks are absent, associated with other characteristics plainly indicating a specific distinction. The Rajah Brooke has confirmed the fact of the existence in the island of Borneo of two distinct species of red orangs; one of a smaller size and somewhat more anthropoid; and the larger species presenting the baboon-like cranium.

In reference to the black chimpanzee of Africa also, another very important addition has been, recently, made to our knowledge of those forms of highly developed quadrumana. In 1847 I received a letter from Dr Savage, a church-missionary at Gaboon, on the west coast of tropical Africa, enclosing sketches of the crania of an ape, which he described as much larger than the chimpanzee, ferocious in its habits, and dreaded by the negro natives more than they dread the lion or any other wild beast of the forest. These sketches showed plainly one cranial characteristic by which the chimpanzee differs in a marked degree from the orangs; viz. that produced by the prominence of the super-orbital ridge, which is wanting in the adult and immature of the orangs. That ridge was strongly marked in the sketches transmitted. At a later period in the same year, were transmitted to me from Bristol two skulls of the same large species of chimpanzee as that notified in Dr Savage's letter; they were obtained from the same locality in Africa, and brought clearly to light evidence of the existence in Africa of a second larger and more powerful ape,—the *Troglodytes gorilla*. They are described and figured in the third Volume of the *Transactions of the Zoological Society*, 1848.

The additional facts, subsequently ascertained respecting the gorilla, although they prove its nearer approach to man than any other tailless ape, have not in any degree affected or invalidated the conclusions at which I then arrived.

Since the date of that memoir, skeletons and the entire carcase preserved in spirits of the gorilla have successively reached the Museums of Paris, Vienna, and London; and have formed the subjects of several memoirs, the results of the recorded observations differing only in regard to the interpretation of the facts.

Dr Wyman, the accomplished anatomical professor at Boston, U.S., agrees with the writer in referring the gorilla to the same genus as the chimpanzee (*Troglodytes*), but he regards the latter as more nearly allied to the human kind.

Professors Duvernoy and Isidore Geoffroy St Hilaire consider the differences in the osteology, dentition, and outward character of the gorilla to be of generic importance; and they enter the species in the zoological catalogues as *Gorilla gina*, the trivial name being that by which the animal is called by the natives of Gaboon; the French naturalists also concur with the American in placing the gorilla below the chimpanzee in the zoological scale; and some have more lately been disposed to place both below the siamangs, gibbons or long-armed apes (*Hylobates*).

The following are the principal external characters of the Gorilla exhibited by the specimen preserved in spirits which was received in 1858, at the British Museum, and is now mounted and exhibited in the Mammalian Gallery. My attention was first attracted by the shortness, almost absence, of neck, due to the backward position of the junction of the head to the trunk, to the great length of the cervical spines, causing the 'nape' to project beyond the 'occiput,' to the great size and elevation of the scapulæ, and to the oblique rising of the clavicles from their sternal attachments to above the level of the angles of the jaw. The brain-case, low and narrow, and the lofty ridges of the skull, make the cranial profile pass in almost a straight line from the occiput to the superorbital ridge, the prominence of which gives the most forbidding feature to the physiognomy of the gorilla; the thick integument overlapping that ridge forming a scowling pent-house over the eyes. The nose is more prominent than in the chimpanzee or orang-utan, not only at its lower expanded part, but at its upper half, where a slight prominence corresponds with that which the author had previously pointed out in the nasal bones. The mouth is very wide, the lips large, of uniform thickness, the upper one with a straight, as if incised margin, not showing the coloured lining membrane when the mouth is shut. The chin is short and receding, the muzzle very prominent. The eyelids with eyelashes, the eyes wider apart than in the orang or chimpanzee; no defined eyebrows; but the hairy scalp continued to the superorbital ridge. The ears are smaller in proportion than in man, much smaller than in the chimpanzee; but the structure of the auricle is more like that of man. On a direct front view of the face, the ears are on the same parallel with the eyes[1]. The huge canines in the male give a most formidable aspect to the beast: they were not fully developed in the younger and entire specimen, now mounted. The profile of the trunk describes a slight convexity from the nape to the sacrum,—there being no inbending at the loins, which seem wanting, the thirteenth pair of ribs being close to the 'labrum ilii.' The chest is of great capacity; the shoulders very wide across; the pectoral regions are slightly marked, and shew a pair of nipples placed as in the chimpanzee and human species. The abdomen is somewhat prominent, both before and at the sides. The pelvis relatively broader than in other apes.

The chief deviations from the human structure are seen in the limbs, which are of great power, the upper ones prodigiously strong. The arm from below the short deltoid prominence preserves its thickness to the condyles; a uniform circumference prevails in the fore-arm; the leg increases in thickness from below the knee to the ankle. There is no 'calf' of the leg. These characters of the limbs are due to the general absence of those partial muscular enlargements which impart the graceful varying curves to the outlines of the limbs in man. Yet they depend rather on excess, than defect, of development of the carneous as compared with the tendinous parts of the limb-muscles, which thus continue of almost the same size from their origin to their insertion, with a proportionate gain of strength to the beast.

The difference in the length of the upper limbs between the gorilla and man is but little in comparison with the trunk; it appears greater through the arrest of development of the lower limbs. Very significant of the closer anthropoid affinities of the gorilla is the superior length of the arm (humerus) to the forearm, as compared with the proportions of those parts in the chim

---

[1] *On the Anthropoid Apes: Proceedings, R. I.* Vol. II. (1855) p. 26, and in the *Transactions of the Zoological Society,* 1848.

panzee. The hair of the arm inclines downward, that of the fore-arm upward, as in the chimpanzee. The thumb extends a little beyond the base of the proximal phalanx of the fore-finger; it does not reach to the end of the metacarpal bone in the chimpanzee or any other ape: the thumb of the siamang is still shorter in proportion to the length of the fingers of the same hand: the philosophical zoologist will see great significance in this fact. In man the thumb extends to, or beyond, the middle of the first phalanx of the fore-finger.

The fore-arm in the gorilla passes into the hand with very slight evidence, by constriction, of the wrist; the circumference of which, without the hair, is fourteen inches, that of a strong man averaging eight inches. The hand is remarkable for its breadth and thickness, and for the great length of the palm, occasioned both by the length of the metacarpus and the greater extent of undivided integument between the digits than in man; these only begin to be free opposite the middle of the proximal or first phalanges in the gorilla. The digits are thus short, and appear as if swollen and gouty; and are conical in shape after the first joint, by tapering to nails, which, being not larger or longer than those of man, are relatively to the fingers much smaller. The circumference of the middle digit at the first joint in the gorilla is $5\frac{1}{2}$ inches; in man, at the same part, it averages $2\frac{3}{4}$ inches. The skin covering the middle phalanx is thick and callous on the backs of the fingers, and there is little outward appearance of the second joint. The habit of the animal to apply those parts to the ground, in occasional progression, is manifested by these callosities. The back of the hand is hairy as far as the divisions of the fingers; the palm is naked and callous. The thumb, besides its shortness, according to the standard of the human hand, is scarcely half so thick as the fore-finger. The nail of the thumb did not extend to the end of that digit; in the fingers the nail projected a little beyond the end, but with a slightly convex worn margin, resembling the human nails in shape, but relatively less.

In the hind-limbs, chiefly noticeable was that first appearance in the quadrumanous series of a muscular development of the gluteus, causing a small buttock to project over each tuber ischii. This structure, with the peculiar expanse (in *Quadrumana*) of the iliac bones, leads to an inference that the gorilla must naturally and with more ease resort occasionally to station and progression on the lower limbs than any other ape.

The same cause as in the arm, viz. a continuance of a large proportion of fleshy fibres to the lower end of the muscles, co-extensive with the thigh, gives a great circumference to that segment of the limb above the knee-joint, and a more uniform size to it than in man. The relative shortness of the thigh, its bone being only eight-ninths the length of the humerus (in man the humerus averages five-sixths the length of the femur), adds to the appearance of its superior relative thickness. Absolutely the thigh is not of greater circumference at its middle than is the same part in man.

The chief difference in the leg, after its relative shortness, is the absence of a 'calf,' due to the non-existence of the partial accumulation of carneous fibres in the gastrocnemii muscles, causing that prominence in the type-races of mankind. In the gorilla the tendo-achillis not only continues to receive the 'penniform' fibres to the heel, but the fleshy parts of the muscles of the foot receive accessions of fibres at the lower third of the leg, to which the greater thickness of that part is due, the proportions in this respect being the reverse of those in man. The leg expands at once into the foot, which has a peculiar and characteristic form, owing to the modifications favouring bipedal motion being super-induced upon an essentially prehensile, quadrumanous type. The heel makes a more decided backward projection than in the chimpanzee; the heel-bone is relatively thicker, deeper, more expanded vertically at its hind end, besides being fully as long as in the chimpanzee. This bone, so characteristic of anthropoid affinities, is shaped and proportioned more like the human calcaneum than in any other ape. The malleoli do not make such well-marked projections as in man; they are marked more by the thickness of the fleshy and tendinous parts of the muscles that pass near them, on their way to be inserted into parts of the foot. Although the foot be articulated to the leg with a slight inversion of the sole, it is more nearly plantigrade than in the chimpanzee or any other ape. The hallux (great toe, thumb of the foot), though not relatively longer than in the chimpanzee, is stronger; the bones are thicker in proportion to their length, especially the last phalanx, which in shape and breadth much resembles that in the human foot. The hallux in its natural position diverges from the other toes at an angle of 60 deg. from the axis of the foot; its base is large, swelling into a kind of ball below, upon which the thick callous epiderm of the sole is continued. The transverse indents and wrinkles show the frequency and freedom of the flexile movements of the two joints of the hallux; the nail is small, flat and short. The sole of the foot gradually expands from the heel forward to the divergence of the hallux, and seems to be here cleft, and almost equally, between the base of the hallux and the common base of the other four digits. These are small and slender in proportion, and their beginnings are enveloped in a common tegumentary sheath as far as the base of the second phalanx. A longitudinal indent at the middle of the sole, bifurcating—one channel defining the ball of the hallux, the other running towards the interspace between the second and third digit,—indicates the action of opposing the whole thumb (which seems rather like an inner lobe or division of the sole), to the outer division terminated by the four short toes. What is termed the 'instep' in man is very high in the gorilla, owing to the thickness of the carneo-tendinous parts of the muscles as they pass from the leg to the foot over this region. The mid-toe (third) is a little longer than the second and fourth; the fifth, as in man, is proportionally shorter than the fourth, and is divided from it by a somewhat deeper cleft. The whole sole is wider than in man—relatively to its length much wider—and in that respect, as well as by the off-set of the hallux, and the definition of its basal ball, more like a hand, but a hand of huge dimensions and of portentous power of grasp.

The hairy integument is continued along the dorsum of the foot to the clefts of the toes, and upon the first phalanx of the hallux: the whole sole is bare.

In regard to the outward coloration of the gorilla, only from the examination of the living animal could the precise shades of colour of the naked parts of the skin be truly described. Much of the epiderm had peeled off the subject of the present description; but fortunately in large patches, and the texture of these had acquired a certain firmness, apparently by the action of the alcohol upon the albuminous basis. The parts of the epiderm remaining upon the face indicated the skin there to be chiefly of a deep leaden hue; it is everywhere finely wrinkled, and was somewhat less dark at the prominent parts of the supraciliary roll and the prominent margins of the nasal 'alæ:' the soles and palms were also of a lighter colour.

Although the general colour of the hair appears, at first sight, and when moist, to be almost black, it is not so, but is rather of a dusky grey: it is decidedly of a less deep tint than in the chimpanzee (*Trogl. niger*): this is due to an admixture of a few reddish, and of more greyish, hairs with the dusky coloured ones which chiefly constitute the 'pelage:' and the above admixture varies at different parts of the body. The reddish hairs are so numerous on the scalp, especially along the upper middle region, as to make their tint rather predominate there; they blend in a less degree with the long hairs upon the sides of the face. The greyish hairs are found mixed with the dusky upon the dorsal, deltoidal and anterior femoral, regions; but on the limbs, not in such proportion as to affect the impression of the general dark colour, at first view. Near the margin of the vent are a few short whitish hairs, as in the chimpanzee. The epiderm of the back shewed the effects of habitual resting, with that part against the trunk or branch of a tree, occasioning the hair to be more or less rubbed off: the epithelium was here very thick and tough.

It is most probable, from the degree of admixture of different coloured hairs above described, that a living gorilla seen in bright sunlight, would in some positions reflect from its surface a colour much more different from that of the chimpanzee than appears by a comparison of the skin of a dead specimen sent home in spirits. It can hardly be doubted, also, that age will make an appreciable difference in the general coloration of the *Troglodytes gorilla.*

The adult male gorilla measures five feet six inches from the sole to the top of the head, the breadth across the shoulders is nearly three feet, the length of the upper limb is three feet four inches, that of the lower limb is two feet four inches; the length of the head and trunk is three feet six inches, whilst the same dimension in man does not average three feet.

In the foregoing remarks are given the results of direct observations made on the first and only entire specimen of the gorilla which has reached England. A more important labour, however, remains. The accurate record of facts in natural history is one and a good aim; the deduction of their true consequences is a better. I proceed, therefore, to reconsider the conclusions from which my experienced French and American fellow-labourers in natural history differ from me.

The first—it may be called the supreme—question in regard to the gorilla is, its place in the scale of nature, and its true and precise affinities.

Is it or not the nearest of kin to human kind? Does it form, like the chimpanzee and orang, a distinct genus in the anthropoid or knuckle-walking group of apes? Are these apes, or are the long-armed gibbons, more nearly related to the genus *Homo?* O the broad-breast-boned quadrumana, are the knuckle-walkers or the brachiators, i.e. the long-armed gibbons, most nearly and essentially related to the human subject?

At the first aspect, whether of the entire animal or of the skeleton, the gorilla strikes the observer as being a much more bestial and brutish animal than the chimpanzee. All the features that relate to the wielding of the strong jaws and large canines are exaggerated; the evidence of brain is less; its proper cavity is more masked by the outgrowth of the strong occipital and other cranial ridges. But then the impression so made that the gorilla is less like man, is the same which is derived from comparing a young with an adult chimpanzee, or some small tailless monkey with a full-grown male orang or chimpanzee. Taking the characters that cause that impression at a first inspection of the gorilla, most of the small South American monkeys are more anthropoid; they have a proportionally larger and more human-shaped cranium, much less prominent jaws, with more equable teeth.

On comparing the skeletons of the adult males of the gorilla, chimpanzee, orang, and gibbon, the globular cranium of the last, and its superior size compared with the jaws and teeth, seemed to shew the gibbons to be more nearly akin to man than any of the larger tailless apes. And this conclusion had been formed by a distinguished French palæontologist, M. Lartet, and accepted by a high geological authority at home[1]. The experienced Professor of Human Anatomy at Amsterdam had been also cited as supporting this view; but I have failed to find any statement of the grounds upon which it was sustained. In the art. *Quadrumana* of Todd's *Cyclopædia*, cited by Lartet[2], Professor Vrolik briefly treats of the osteology of the *Quadrumana* according to their natural families. In 'a first genus, *Simia* proper, or ape,' he includes the chimpanzee or orang, noticing some of the chief points by which these apes approach the nearest to man. He next goes to the second genus, the gibbon (*Hylobates*), notices their ischial callosities, and the nearer approach of their molars, in their rounded form, to the teeth of *carnivora* than the molars-of the genus *Simia*. Then, comparing the siamang with other species of *Hylobates*, Vrolik says, 'its skeleton approaches most to that of man,' which may be true

[1] Sir C. Lyell, Supplement to the 5th Edition of a *Manual of Elementary Geology*, 1859, p. 15.

[2] *Comptes Rendus de l'Académie des Sciences*, Juillet 28, 1856.

in comparison with other gibbons, but certainly is not so as respects the higher *Simiæ*. No details are given to illustrate the proposition even in its more limited application; but the minor length of the arms in the siamang, as compared with *Hylobates lar*, was probably the obvious character in Vrolik's mind.

The appearance of superior cerebral development in the siamang and other long-armed apes is due to their small size and the concomitant feeble development of their jaws and teeth. The same appearance makes the small platyrrhine monkeys of South America equally anthropoid in their facial physiognomy, and much more human-like than are the great orangs and chimpanzees. It is an appearance which depends upon the precocious growth of the brain, as dependent on the law of its development. In all *quadrumana* the brain has reached its full size before the second set of teeth is acquired, almost before the first set is shed. If a young gorilla, chimpanzee, or orang, be compared with a young siamang, of corresponding age, the absolutely larger size and better shape of brain, the deeper and more numerous convolutions of the cerebrum, and the more completely covered cerebellum, unequivocally demonstrate the higher organization of the shorter-armed apes; 'in the structure of the brain,' writes Vrolik[1], in accordance with all other comparative anatomists, 'they' (chimpanzee and orang-utan) 'approach the nearest to man.' The degree to which the chimpanzee and orang so resembled the human type seemed much closer to Cuvier, who knew those great apes only in their immaturity, with their small milk-teeth and precociously developed brain. Accordingly, the anthropoid characters of the *Simia satyrus* and *Simia troglodytes*, as deduced from the facial angle and dentition, are proportionally exaggerated in the *Règne Animal*[2]. As growth proceeds, the milk-teeth are shed, the jaws expand, the great canines succeed their diminutive representatives, the biting muscles gain a proportional increase of carneous fibres, their bony fulcra respond to the call for increased surface of attachment, and the sagittal and occipital crests begin to rise : but the brain grows no more ; its cranial box retains the size it shewed in immaturity; it finally becomes masked by the superinduced osseous developments in those apes which attain the largest stature and wield the most formidably armed jaws. Yet under this disguise of physical force, the brain is still the better and the larger than is that of the little long-armed ape,

[1] Art. *Quadrumana, Cyclopædia of Anatomy*, Vol. IV. p. 195.
[2] Ed. 1829, pp. 87, 89.

which retains throughout life so much more of the characters of immaturity, especially in the structure of the skull.

The siamang and other gibbons have smaller lower but longer upper canines, relatively, than in the orangs and chimpanzees; the permanent ones more quickly attain their full size, and are sooner in their place in the jaws; consequently the last molar teeth—what we call the 'wisdom-teeth'—come last into place as they do in the human species. But, if this be interpreted as of importance in determining the relative affinity of the longer-armed and shorter-armed apes to man, it is a character in which, as in their seeming superior cerebral development, the *Hylobates* agree with some much lower *Quadrumana* with still smaller canines. The comparative anatomist, pursuing this most interesting comparison with clear knowledge of the true conditions and significance of a globular cranium and small jaws within the quadrumanous order, turns his attention to the true distinctive characters of the human organization.

In respect to the brain, he would look not so much for its relative size to the body, as for its relative size in the species compared one with another in the same natural group. He would inquire what quadrumanous animal shews absolutely the biggest brain? what species shows the deepest and most numerous and winding convolutions? in which is the cerebrum largest, as compared with the cerebellum? If he found all these characters highest in the gorilla, he would not be diverted from the just inference because the great size and surpassing physical power attained in that species masked the true data from obvious view.

The comparative anatomist would look to the cæcum and the ischial integument: if he found in one subject of his comparisons (*Troglodytes*) a long 'appendix vermiformis cæci,' as in man, but no 'callosities,' and in another subject (*Hylobates*) the ischial callosities but only a short rudiment of the cæcal appendix, he would know which of the two tailless apes were to be placed next 'the monkeys with ischial callosities and no vermiform appendix,' and which formed the closer link toward man. He would find that the anthropoid intestinal and dermal characters were associated with the absolutely larger and better developed brain in the gorilla, chimpanzee, and orang; whilst the lower quadrumanous characters exhibited by the cæcum and nates were exhibited by the smaller-brained and longer-armed tailless gibbons.

Pursuing the comparison through the complexities of the bony framework, the comparative anatomist would first glance at the more obvious characters; and such, indeed, as would be given by the entire animal. The characteristics of the limbs in man are their near equality of length, but the lower limbs are the longest. The arms in man reach to below the middle of the thigh; in the gorilla they nearly attain the knee; in the chimpanzee they reach below the knee; in the orang they reach the ankle; in the siamang they reach the sole; in most gibbons the whole palm can be applied to the ground without the trunk being bent forward beyond its naturally inclined position on the legs. These gradational differences coincide with other characters determining the relative proximity of the apes compared with man. In no quadrumana does the humerus exceed the ulna so much in length as in man; only in the very highest and most anthropoid, viz. the gorilla and chimpanzee, does it exceed the ulna at all in length; in all the rest, as in the lower quadrupeds, the fore-arm is longer than the arm.

The humerus, in the gorilla, though less long, compared with the ulna, than in man, is longer than in the chimpanzee; in the orang it is shorter than the ulna; in the siamang and other gibbons it is much shorter, the peculiar length of arm in those 'long-armed apes' is chiefly due to the excessive length of the antibrachial bones.

The difference in the length of the upper limbs, as compared with the trunk, is but little between man and the gorilla. The elbow-joint in the gorilla, as the arm hangs down, is opposite the 'labrum ilii,' the wrist opposite the 'tuber ischii;' it is rather lower down in the chimpanzee; is opposite the knee-joint in the orang; and opposite the ankle-joint in the siamang.

Man's perfect hand is one of his peculiar physical characters; that perfection is mainly due to the extreme differentiation of the first from the other four digits, and its concomitant power of opposing them as a perfect thumb. An opposable thumb is present in the hand of most *Quadrumana*, but is usually a small appendage compared with that of man. It is relatively largest in the gorilla. In this ape the thumb reaches to a little beyond the base of the first phalanx of the fore-finger; it does not reach to the end of the metacarpal bone of the fore-finger in the chimpanzee, orang, or gibbon; it is relatively smallest in the last tailless ape. In man the thumb extends to or beyond the middle of the first phalanx of the fore-finger. The philosophical zoologist will see great significance in the results of this comparison. Only in the gorilla and chimpanzee are the carpal bones eight in number, as in man; in the orangs and gibbons they are nine in number, as in the tailed monkeys.

The scapulæ are broader in the gorilla than in the chimpanzee, orang, or long-armed apes; they come nearer to the proportions of that bone in man. But a more decisive resemblance to the human structure is presented by the iliac bones. In no other ape than the gorilla do they bend forward, so as to produce a pelvic concavity; nor are they so broad in proportion to their length in any ape as in the gorilla. In both the chimpanzee and orang the iliac bones are flat, or present a concavity rather at the back than at the forepart. In the siamang they are not only flat, but are narrower and longer, resembling the iliac bones of tailed monkeys and ordinary quadrupeds.

The lower limbs, though characteristically short in the gorilla, are longer in proportion to the upper limbs, and also to the entire trunk, than in the chimpanzee; they are much longer in both proportions and more robust than in the orangs or gibbons. But the guiding points of comparisons here are the heel and the hallux (great toe or thumb of the foot).

The heel in the gorilla makes a more decided backward projection than in the chimpanzee; the heel-bone is relatively thicker, deeper, more expanded vertically at its hind end, besides being fully as long as in the chimpanzee: it is in the gorilla shaped and proportioned more like the human calcaneum than in any other ape. Among all the tailless apes the calcaneum in the siamang and other gibbons least resembles in its shape or proportional size that of man.

Although the foot be articulated to the leg with a slight inversion of the sole it is more nearly plantigrade in the gorilla than in the chimpanzee. The orang departs far, and the gibbons farther, from the human type in the inverted position of the foot.

The great toe which forms the fulcrum in standing or walking is perhaps the most characteristic peculiarity in the human structure; it is that modification which differentiates the foot from the hand, and gives the character to his order (*Bimana*). In the degree of its approach to this development of the hallux the quadrumanous animal makes a true step in affinity to man.

The orang-utan and the siamang, tried by this test, descend far and abruptly below the chimpanzee and gorilla in the scale. In the orang the hallux does not reach to the end of the metacarpal of the second toe; in the chimpanzee and gorilla it reaches to the end of the first phalanx of the second toe; but in the gorilla the hallux is thicker and stronger than in the chimpanzee. In both, however, it is a true thumb, by position, diverging from the other toes, in the gorilla, at an angle of 60° from the axis of the foot.

Man has 12 pairs of ribs, the gorilla and chimpanzee have 13 pairs, the orangs have 12 pairs, the gibbons have 13 pairs. Were the naturalist to trust to this single character, as some have trusted to the cranio-facial one, and in equal ignorance of the real condition and value of both, he might think'that the orangs (*Pithecus*) were nearer akin to man than the chimpanzees (*Troglodytes*) are. But man has sometimes a thirteenth pair of ribs ; and what we term ' ribs' are but vertebral elements or appendages common to nearly all the true vertebræ in man, and only so called, when they become long and free. The genera *Homo*, *Troglodytes*, and *Pithecus*, have precisely the same number of vertebræ : if *Troglodytes*, by the development and mobility of the pleurapophyses of the 20th vertebræ from the occiput seem to have an additional thoracic vertebra, it has one vertebra less in the lumbar region. So, if there be, as has been observed in the same genus, a difference in the number of sacral vertebræ, it is merely due to a last lumbar having coalesced with what we reckon the first sacral vertebra in man.

The thirteen pairs of ribs, therefore, in the gorilla and chimpanzee are of no weight, as against the really important characters significative of affinity with the human type. But, supposing the fact of any real value, how do the advocates of the superior resemblance of the gibbon's skeleton to that of man dispose of the thirteenth pair of ribs ?

In applying the characters of the skull to the determination of the important question at issue those had first to be ascertained by which the genus *Homo* trenchantly differs from the genus *Simia*, of Linnæus. To determine these osteal distinctions I have compared the skulls of many individuals of different varieties of the human race together with those of the male, female, and young of species of *Troglodytes*, *Pithecus*, and *Hylobates;* the detailed results of which comparisons will be found in the Catalogue of the Osteological Series in the Museum of the Royal College of Surgeons, 4to, 1853. In the present Appendix, I restrict myself to a few of these results.

The first and most obvious differential character is the globular form of the brain-case, and its superior relative size to the face, especially the jaws, in man. But this, for the reasons already assigned, is not an instructive or decisive character, when comparing quadrumanous species, in reference to the question at issue. It is exaggerated in the human child, owing to the acquisition of its full, or nearly full size, by the brain, before the jaws have expanded to lodge the second set of teeth. It is an anthropoid character in which the quadrumana resemble man in proportion to the diminution of their general bulk. If a gorilla, with milk-teeth, have a somewhat larger brain and brain-case than a chimpanzee at the same immature age, the acquisition of greater bulk by the gorilla, and of a more formidable physical development of the skull, in reference to the great canines in the male, will give to the chimpanzee the appearance of a more anthropoid character, which really does not belong to it; which could be as little depended upon in a question of precise affinity as the like more anthropoid characters of the female, as compared with the male, gorilla or chimpanzee.

Much more important and significant are the following characters of the human skull :—the position and plane of the occipital foramen; the proportion and size of the condyloid and petrous processes; the mastoid processes, which relate to balancing the head upon the trunk in the erect attitude; the small premaxillaries and concomitant small size of the incisor teeth, as compared with the molar teeth. These characters relate to the superiority of the psychical over the physical powers in man. They govern the feature in which man recedes from the brute; and to them may be added the prominence of the nasal bones in most, and in all the typical, races of man. The somewhat angular form of the bony orbits, tending to a square, with the corners rounded off, is, likewise, a good human character of the skull; which is difficult to comprehend as an adaptive one, and therefore the better in the present inquiry. The same may be said of the production of the floor of the tympanic or auditory tube into the plate called 'vaginal.'

Believing the foregoing to be sufficient to test the respective degrees of affinity to man within the limited group of quadrumana to which it is now proposed to apply them, I forbear to cite the characters of minor importance. The question at issue is, as between the anthropoid apes and man. Cuvier deemed the orang (*Pithecus*) to be nearer akin to man than the chimpanzee (*Troglodytes*) is. That belief has long ceased to be entertained. I proceed, therefore, to compare the gorilla, chimpanzee, and gibbon, in reference to their human affinities.

Most naturalists entering upon this question would first look to the premaxillary bones, or, owing to the early confluence of those bones with the maxillaries in the gorilla and chimpanzee, to the part of the upper jaw containing the incisive teeth, on the development of which depends the prognathic or brutish character of a skull. Now the extent of the premaxillaries below the nostril is not only relatively but absolutely less in the gorilla, and consequently the profile of the skull is less convex at this part, or less 'prognathic,' than in the chimpanzee. Notwithstanding the degree in which the skull of the gorilla surpasses in size that of the chimpanzee, especially when the two are compared on a front view, the breadth of the premaxillaries and of the four incisive teeth is the same in both. In the relative degree, therefore, in which these bones are smaller than in the chimpanzee, the gorilla, in this most important character, comes nearer to man. In the gibbons the incisors are relatively smaller than in the gorilla, but the premaxillaries bear the same proportional size, in the adult male siamang.

Next, as regards the nasal bones. In the chimpanzee, as in the orangs and gibbons, they are as flat to the face as in any of the lower *Simiæ*. In the gorilla, the median coalesced margins of the upper half of the nasal bones are produced forwards; in a slight degree it is true, but affording a most significant evidence of nearer resemblance to man. In the same degree they impress that anthropic feature upon the face of the living gorilla. In some pig-faced baboons there are ridges and prominences in the naso-facial part of the skull; but they do not really affect the question as between the gorilla and chimpanzee. All naturalists know that the semnopithèques of Borneo have long noses; but the proboscidiform appendage which gives so ludicrous a mask to those monkeys is scarcely the homologue of the human nose, and is unaccompanied by any such modification of the nose-bones as gives the true anthropoid character to the human skull, and to which only the gorilla, in the ape tribe, makes any approximation.

No orang, chimpanzee, or gibbon shews any rudiment of mastoid processes; but they are present in the gorilla, smaller indeed than in man, but unmistakeable; they are, as in man, cellular, and with a thin outer plate of bone. This fact led me to express, when in respect to the gorilla, only the skull had reached me, the following inference, viz.: 'from the nearer approach which the gorilla makes to man in comparison with the chimpanzee, or orang, in regard to the mastoid processes, that it assumed more nearly and more habitually the upright attitude than those inferior anthropoid apes do.' This inference has been fully borne out by the rest of the skeleton of the gorilla, subsequently acquired.

In the chimpanzee, as in the orangs, gibbons, and inferior *Simiæ*, the lower surface of the long tympanic or auditory process is more or less flat and smooth, developing in the chimpanzee only a slight tubercle, anterior to the stylohyal pit. In the gorilla the auditory process is more or less convex below, and developes a ridge, answering to the vaginal process, on the outer side of the carotid canal. The processes posterior and internal to the glenoid articular surface, are better developed, especially the internal one, in the gorilla than in the chimpanzee; the ridge which extends from the ectopterygoid along the inner border of the foramen ovale, terminates in the gorilla by an angle or process answering to that called 'styliform' or 'spinous' in man, but of which there is no trace in the chimpanzee, orang, or gibbon.

The orbits have a full oval form in the orang; they are almost circular in the chimpanzee and siamang; more nearly circular, and with a more prominent rim in the smaller gibbons; in the gorilla alone do they present the form which used to be deemed peculiar to man. There is not much physiological significance in some of the latter characters; but, on that very account, I deem them more instructive and guiding in the actual comparison. The occipital foramen is nearer the back part of the cranium, and its plane is more sloping, less horizontal, in the siamang, than in the chimpanzee and gorilla. Considering the less relative prominence of the fore part of the jaws in the siamang, as compared with the chimpanzee, the occipital character of that gibbon and of other species of *Hylobates* indicates well their inferior position in the quadrumanous scale.

In the greater relative size of the molars, compared with the incisors, the gorilla makes an important closer step towards man than does the chimpanzee. The molar teeth are relatively so small in the siamang, that notwithstanding the small size of the incisors, the proportion of those teeth to the molars is only the same as in the gorilla: in other gibbons (*Hylobates lar*), the four lower incisors occupy an extent equal to that of the first four molars, in the chimpanzee equal to that of the first three molars, in the siamang equal to that of the first two molars and rather more than half of the third, in man equal to the first two molars and half of the third: in this comparison the term molar is applied to the bicuspids.

The proportion of the ascending ramus to the length of the lower jaw tests the relative affinity of the tailless apes to man.

In a profile of the lower jaw, compare the line drawn vertically from the top of the coronoid process to the horizontal length along the alveoli. In man and the gorilla it is about 7-10ths, in the chimpanzee 6-10ths, in the siamang it is only 4-10ths. The siamang further differs in the shape and production of the angle of the jaw, and in the shape of the coronoid process, approaching the lower simiæ in both these characters. In the size of the post-glenoid process, in the shape of the glenoid cavity which is almost flat, in the proportional size of the petrous bone, and in the position of the foramen caroticum, the siamang departs further from the human type and approaches nearer that of the tailed simiæ than the gorilla does, and in a marked degree.

Every legitimate deduction from a comparison of cranial characters makes the tailless *Quadrumana* recede from the human type in the following order,—gorilla, chimpanzee, orangs, gibbons; and the last-named in a greater and more decided degree.

Those comparisons have of late been invested with additional interest from the discoveries of remains of quadrumanous species in different members of the tertiary formations.

The first quadrumanous fossil, the discovery of which by Lieuts. Baker and Durand is recorded in the *Journal of the Asiatic Society of Bengal*, for November, 1836, has proved to belong, like subsequently discovered quadrumanous fossils in the Sewalik (probably miocene) tertiaries, to the Indian genus *Semnopithecus*. The quadrumanous fossils discovered in 1839, in the eocene deposits of Suffolk, belong to a genus (*Eopithecus*) having its nearest affinities with *Macacus*. The monkey's molar tooth from the pliocene beds of Essex is most closely allied to the *Macacus sinicus*. The remains of the large monkey, 4 feet in height, discovered in 1839 by Dr Lund in a limestone cavern in Brazil was shewn by its molar dentition $\left(p\, \dfrac{3-3}{3-3},\ m\, \dfrac{3-3}{3-3} \right)$ to belong to the platyrrhine family now peculiar to South America. The lower jaw and teeth of the small quadrumane discovered by M. Lartet in a miocene bed of the south of France, and described by him and De Blainville, is so closely allied to the gibbons as to scarcely justify the generic separation which has been made for it under the name *Pliopithecus*.

Finally, a portion of a lower jaw with teeth and the shaft of a humerus of a quadrumanous animal (*Dryopithecus*), equalling the size of those bones in man, have been discovered by M. Fontan, of Saint-Gaudens, in a marly bed of upper miocene age, forming the base of the plateau on which that town is built. The molar teeth present the type of grinding surface of those of the gibbons (*Hylobates*), and as in that genus the second true molar is larger than the first, not of equal size, as in the human subject and chimpanzee. The premolars have a greater antero-posterior extent, relatively, than in the chimpanzee; and in this respect agree more with those in the siamang. The first premolar has the outer cusp raised to double the height of that of the second; its inner lobe appears from M. Lartet's figure to be less developed than in the gorilla, certainly less than in the chimpanzee. The posterior talon of the second premolar is more developed, and consequently the fore and aft extent of the tooth is greater than in the chimpanzee; thereby the second premolar of *Dryopithecus* more resembles that in *Hylobates*, and departs further from the human type.

The canine, judging from the figures published by M. Lartet[1], seems to be less developed than in the male chimpanzee, gorilla, or orang. In which character the fossil, if it belonged to a male, makes a nearer approach to the human type; but it is one which many of the inferior monkeys also exhibit, and is by no means to be trusted as significant of true affinity, supposing even the sex of the fossil to be known as being male.

The shaft of the humerus, found with the jaw, is peculiarly rounded, as it is in the gibbons and sloths, and offers none of those angularities and ridges which make the same bone in the chimpanzee and orang come so much nearer in shape to the humerus of the human subject. The fore part of the jaw, as in the siamang, is more nearly vertical than in the gorilla or chimpanzee, but whether the back part of the jaw may not have departed in a greater degree from the human type than the fore part approaches it, as is the case in the siamang, the state of the fossil does not allow of determining. One significant character is, however, present,—the shape of the fore part of the coronoid process. It is slightly convex forwards, which causes the angle it forms with the alveolar border to be less open. The same character is present in the gibbons. The fore part of the lower half of the coronoid process in man is concave, as it is likewise in the gorilla and chimpanzee. I am acquainted with this interesting fossil, referred to a genus

[1] *Comptes Rendus de l'Académie des Sciences*, Paris, Vol. XLIII.

called *Dryopithecus*, only by the figures published in the 43rd volume of the *Comptes Rendus de l'Académie des Sciences*. From these it appears that the canine, two premolars, and first and second true molars are in place. The socket of the third molar is empty, but widely open above; from which I conclude that the third molar had also cut the gum, the crown being completed, but not the fangs. If the last molar had existed as a mere germ, it would have been preserved in the substance of the jaw.

In a young siamang, with the points of the permanent canines just protruding from the socket, the crown of the last molar is complete, and on a level with the base of that of the penultimate molar, whence I infer that the last molar would have cut the gum as soon as, if not before, the crown of the canine had been completely extricated. This dental character, the conformation and relative size of the grinding teeth, especially the fore-and-aft extent of the premolars, all indicate the close affinity of the *Dryopithecus* with the *Pliopithecus* and existing gibbons; and this, the sole legitimate deduction from the maxillary and dental fossils, is corroborated by the fossil humerus, fig. 9, in the above-cited plate.

There is no law of correlation by which, from the portion of jaw with teeth of the *Dryopithecus*, can be deduced the shape of the nasal bones and orbits, the position and plane of the occipital foramen, the presence of mastoid and vaginal processes, or other cranial characters determinative of affinity to man; much less any ground for inferring the proportions of the upper to the lower limbs, of the humerus to the ulna, of the pollex to the manus, or the shape and development of the iliac bones. All those characters which do determine the closer resemblance and affinity of the genus *Troglodytes* to man, and of the genus *Hylobates* to the tailed monkeys, are at present unknown in respect of the *Dryopithecus*. A glance at fig. 5 (*Gorilla*), and fig. 7 (*Dryopithecus*), of the plate of M. Lartet's memoir, would suffice to teach their difference of bulk, the gorilla being fully one-third larger. The statement that the parts of the skeleton of the *Dryopithecus* as yet known, viz., the two branches of the lower jaw and the humerus, 'are sufficient to shew that in anatomical structure, as well as stature, it came nearer to man than any quadrumanous species, living or fossil, before known to zoologists[1],' is without the support of any ade-

---

[1] Sir Chas. Lyell, Supplement to the Fifth Edition of a *Manual of Elementary Geology*, 8vo., 1859, p. 14.

quate fact, and in contravention of most of those to be deduced from M. Lartet's figures of the fossils. Those parts of the *Dryopithecus* merely shew—and the humerus in a striking manner—its nearer approach to the gibbons. The most probable conjecture being that it bore to them, in regard to size, the like relations which Dr Lund's *Protopithecus* bore to the existing *Mycetes.* Whether, therefore, strata of such high antiquity as the miocene may reveal to us 'forms in any degree intermediate between the chimpanzee and man' awaits an answer from discoveries yet to be made; and the anticipation that the fossil world 'may hereafter supply new osteological links between man and the highest known quadrumana' may be kept in abeyance until that world has furnished us with the proofs that a species did formerly exist which came as near to man as does the orang, the chimpanzee, or the gorilla.

Of the nature and habits of the last-named species, which really offers the nearest approach to man of any known ape, recent or fossil, the lecturer had received many statements from individuals resident at or visitors to the Gaboon, from which he selected the following as most probable, or least questionable.

Gorilla-land is a richly wooded extent of the western part of Africa, traversed by the rivers Danger and Gaboon, and extending from the equator to the 10th or 15th degree of south latitude. The part where the gorilla has been most frequently met with presents a succession of hill and dale, the heights crowned with lofty trees, the valleys covered by coarse grass, with partial scrub or scattered shrubs. Fruit trees of various kinds abound both on the hills and in the valleys; some that are crude and uncared for by the negroes are sought out and greedily eaten by the gorillas, and as different kinds come to maturity at different seasons, they afford the great denizen of the woods a successive and unfailing supply of these indigenous fruit trees. I am able through the contributions of kind and zealous correspondents to specify the following:—

The palm-nut (*Elais guiniensis*) of which the gorillas greatly affect the fruit and upper part of the stipe, called the 'cabbage.' The negroes of the Gaboon have a tradition that their forefathers first learnt to eat the 'cabbage,' from seeing the gorilla eat it, concluding that what was good for him must be good for man.

The 'ginger-bread tree' (*Parinarium excelsum*), which bears a plum-like fruit.

The papau tree (*Carica papaya*).

The banana (*Musa sapientium*), and another species (*Musa paradisiaca*).

The *Amomum Afzelii* and *Am. grandiflorum*.

A tree, with a shelled fruit, like a walnut, which the gorilla breaks open with the blow of a stone.

A tree, also botanically unknown, with a fruit like a cherry.

Such fruits and other rich and nutritious productions of the vegetable kingdom, constitute the staple food of the gorilla, as they do of the chimpanzee. The molar teeth, which alone truly indicate the diet of an animal, accord with the statements as to the frugivorous character of the gorilla: but they also sufficiently answer to an omnivorous habit to suggest that the eggs and callow brood of nests discovered in the trees frequented by the gorilla might not be unacceptable.

The gorilla makes a sleeping place like a hammock, connecting the branches of a sheltered and thickly leaved part of a tree by means of the long tough slender stems of parasitic plants, and lining it with the broad dried fronds of palms, or with long grass. This hammock-like abode may be seen at different heights, from 10 feet to 40 feet from the ground, but there is never more than one such nest in a tree.

They avoid the abodes of man, but are most commonly seen in the months of September, October, and November, after the negroes have gathered their outlying rice crops, and have returned from the 'bush' to the village. So observed, they are described to be usually in pairs; or, if more, the addition consists of a few young ones, of different ages, and apparently of one family. The gorilla is not gregarious. The parents may be seen sitting on a branch, resting the back against the tree-trunk—the hair being generally rubbed off the back of the old gorilla from that habit—perhaps munching their fruits, whilst the young gorillas are at play, leaping and swinging from branch to branch, with hoots or harsh cries of boisterous mirth.

If the old male be seen alone, or when in quest of food, he is usually armed with a stout stick, which the negroes aver to be the weapon with which he attacks his chief enemy the elephant. Not that the elephant directly or intentionally injures the gorilla, but, deriving its subsistence from the same substances, the ape regards the great proboscidian as a hostile intruder. When therefore he discerns the elephant pulling down and wrenching off the branches of a favourite tree, the gorilla, stealing along the bough, strikes the sensitive proboscis of the elephant with a violent blow of his club, and drives off the startled giant trumpeting shrilly with rage and pain.

In passing along the ground from one detached tree to another the gorilla is said to walk semi-erect, with the aid of his club, but with a waddling awkward gait; when without a stick, he has been seen to walk as a biped, with his hands clasped across the back of his head, instinctively so counterpoising its forward projection. If the gorilla be surprised and approached while on the ground, he drops his stick, betakes himself to all-fours, applying the back part of the bent knuckles of his fore-hands to the ground, and makes his way rapidly, with an oblique swinging kind of gallop, to the nearest tree. There he awaits his pursuer, especially if his family be near, and requiring his defence. No negro willingly approaches the tree in which the male gorilla keeps guard. Even with a gun the negro does not rashly make the attack, but reserves his fire in self-defence. The enmity of the gorilla to the whole negro race, male and female, is uniformly testified to. The young men of the Gaboon tribe make armed excursions into the forests, in quest of ivory. The enemy they most dread on these occasions is the gorilla. If they have come unawares too near him with his family, he does not, like the lion, sulkily retreat, but comes rapidly to the attack, swinging down to the lower branches, and clutching at the nearest foe. The hideous aspect of the animal, with his green eyes flashing with rage, is heightened by the skin over the prominent roof of the orbits being drawn rapidly backward and forward, the hair erected, and causing a horrible and fiendish scowl. If fired at and not mortally hit, the gorilla closes at once upon his assailant and inflicts most dangerous, if not deadly, wounds with his sharp and powerful tusks. The commander of a Bristol trader informed me that he had seen a negro at the Gaboon frightfully mutilated by the bite of the gorilla, from which he had recovered. Another negro exhibited to the same voyager a gun-barrel bent and partly flattened by the bite of a wounded gorilla, in its death-struggle.

Negroes when stealing through the gloomy shades of the tropical forest become sometimes aware of the proximity of one of these frightfully formidable apes by the sudden disappearance of one of their companions, who is hoisted up into the tree, uttering, perhaps, a short choking cry. In a few minutes he falls to the ground a strangled corpse. The gorilla, watching his opportunity, has let down his huge hind-hand, seized the passing negro by the neck, with vice-like grip, has drawn him up to higher branches, and dropped him when his struggles had ceased.

The strength of the gorilla is such as to make him a match for a lion, whose tusks his own almost rival. Over the leopard, invading the lower branches of the gorilla's dwelling tree, he will gain an easier victory; and the huge canines, with which only the male gorilla is furnished, doubtless have been assigned to him for defending his mate and offspring.

The skeleton of the old male gorilla obtained for the British Museum in 1857, shews an extensive fracture, badly united, of the left arm-bone, which has been shortened, and gives evidence of long suffering from abscess and partial exfoliation of bone. The upper canines have been wrenched out or shed, some time before death, for their sockets have become absorbed.

The redeeming quality in this fragmentary history of the gorilla is the male's care of his family, and the female's devotion to her young.

It is reported that a French natural-history collector, accompanying a party of the Gaboon negroes into the gorilla woods, surprised a female with two young ones on a large boabdad (*Adansonia*), which stood some distance from the nearest clump. She descended the tree, with the youngest clinging to her neck, and made off rapidly on all-fours to the forest, and escaped. The deserted young one on seeing the approach of the men began to utter piercing cries: the mother, having disposed of her infant in safety, returned to rescue the older offspring, but before she could descend with it her retreat was cut off. Seeing one of the negroes level his musket at her, she, clasping her young with one arm, waved the other, as if deprecating the shot; the ball passed through her heart, and she fell with her young one clinging to her. It was a male, and survived the voyage to Havre, where it died on arriving. I have examined the skeleton of this young gorilla in the museum of natural history at Caen, and am indebted to Professor Deslongchamps, Dean of the Faculty of Sciences in that town, for drawings of this rare specimen.

There might be more difficulty in obtaining a young gorilla for exhibition than a young chimpanzee. But as no full-grown chimpanzee has ever been captured, we cannot expect the larger and much more powerful adult gorilla to be ever taken alive.

A bold negro, the leader of an elephant-hunting expedition,

being offered a hundred dollars if he would bring back a live gorilla, replied, 'If you gave me the weight of yonder hill in gold coins, I could not do it!'

All the terms of the aborigines in respect to the gorilla imply their opinion of his close kinship to themselves. But they have a low opinion of his intelligence. They say that during the rainy season he builds a house without a roof. The natives on their hunting excursions light fires for their comfort and protection by night; when they have gone away, they affirm that the gorilla will come down and warm himself at the smouldering embers, but has not wit enough to throw on more wood, out of the surrounding abundance, to keep the fire burning,—'the stupid old man!'

Every account of the habits of a wild animal obtained at second hand from the reports of aborigines has its proportion of 'apocrypha.' I have restricted myself to the statements that have most probability and are in accordance with the ascertained structures and powers of the animal, and would only add the averment and belief of the Gaboon negroes that when a gorilla dies, his fellows cover the corpse with a heap of leaves and loose earth collected and scraped up for the purpose.

A most singular phenomenon in natural history, if one reflects on the relations of things, is this gorilla! Limited as it is in its numbers and geographical range, one discerns that the very peculiar conditions of its existence—abundance of wild fruit—needs must be restricted in space; but, concurring in a certain part of Africa, there lives the creature to enjoy them.

The like conditions exist in Borneo and Sumatra, and there also a correlative human-like ape, of similar stature, tooth-armour, and force, exists at their expense. Neither orangs nor gorillas, however, minister to man's use directly or indirectly. Were they to become extinct, no sign of the change or break in the links of life would remain. What may be their real significance?

In regard to the ancient notices which may relate to the great anthropoid ape of Africa, I may quote the following passage from the 'Periplus,' or Voyage of Hanno, which has been supposed to refer to the species in question:—'On the third day, having sailed from thence, passing the streams of fire, we came to a bay called the Horn of the South. In the recess there was an island like the first, having a lake, and in this there was another island full of wild men. But much the greater part of them were women, with hairy bodies, whom the interpreters called "gorillas." But, pursuing them, we were not able to take the men; they all escaped, being able to climb the precipices, and defended themselves with pieces of rock. But three females, who bit and scratched those who led them, were not willing to follow. However, having killed them, we flayed them, and conveyed the skins to Carthage. For we did not sail any further, as provisions began to fail.' This encounter indicates, therefore, the southernmost point on the west coast of Africa reached by the Carthaginian navigator.

To an inquiry by an eminent Greek scholar, how far the newly-discovered great ape of Africa bore upon the question of the authenticity of the Periplus? I have replied:—'The size and form of the great ape, now called "gorilla," would suggest to Hanno and his crew no other idea of its nature than that of a kind of human being; but the climbing faculty, the hairy body, and the skinning of the dead specimens, strongly suggest that they were large anthropoid apes. The fact that such apes, having the closest observed resemblance to the negro, being of human stature and with hairy bodies, do still exist on the west coast of Africa, renders it highly probable that such were the creatures which Hanno saw, captured, and called "Gorullai." '

The brief observation made by Battell in West tropical Africa, 1590, recorded in Purchas's *Pilgrimages, or Relations of the World*, 1748, of the nature and habits of the large human-like ape which he calls ' pongo,' more decidedly refers to the gorilla. Other notices, as by Nieremberg and Bosman, applied by Buffon to Battell's pongo, were deemed valueless by Cuvier, who altogether rejected the conclusions of his great predecessor as to the existence of any such ape. ' This name of pongo or boggo, given in Africa to the chimpanzee or to the mandril, has been applied,' writes Cuvier, ' by Buffon to a pretended great species of ourang-utan, which was nothing more than the imaginary product of his combinations.' After the publication of Cuvier's *Règne Animal*, the supposed species was, by the high authority of its author, banished from natural history; it has only been authentically reintroduced since the intelligent attention of Dr Savage was directed to the skull, which he first saw at the Gaboon in 1847, and took my opinion upon.

Having premised the foregoing account of the mature characters of the different species of orangs and chimpanzees, in regard to their relative proximity to the human species, I next proceed to shew how their structure contrasts with that of man. With regard to the dentition of these anthropoid apes, the number and kinds of the teeth, like those of all the quadrumana of the old world, correspond with those in the human subject; but all these apes differ in the larger proportionate size of the canine teeth, which necessitates a certain break in the series, in order that the prolonged points of the canine teeth may pass into their place when the mouth is completely closed. In addition to the larger proportionate size of the incisors and canines, the bicuspids in both jaws are implanted by three distinct fangs—two external and one internal: in the human species, the bicuspids are implanted by one external and one internal fang : in the highest races of man these two fangs are often connate ; very rarely is the external fang divided, as it constantly is in all the species of the orang and the chimpanzee.

With regard to the catarrhine, or old-world quadrumana, the number of milk teeth is twenty, as in the human subject. But both chimpanzees and orangs differ from man in the order of development of the permanent series of teeth : the second true molar comes into place before either of the bicuspids have cut the gum, and the last molar is acquired before the permanent canine. We may well suppose that the larger grinders are earlier required by the frugivorous apes than by the omnivorous human race ; and one condition of the earlier development of the canines and bicuspids in man, may be their smaller relative size as compared with the apes. The great difference is the predominant development of the permanent canine teeth, at least in the males of the orangs and chimpanzees; for this is a sexual distinction, the canines in the females never presenting the same large proportion. In man, the dental system, although the formula is the same as in the apes, is peculiar for the equal length of the teeth, arranged in an uninterrupted series, and shews no sexual distinctions. The characteristics of man are exhibited in a still more important degree in the parts of the skeleton. His whole framework proclaims his destiny to carry himself erect ; the anterior extremities are liberated from any service in the mere act of locomotion.

With regard to the foot, I have shewn in my work *On the Nature of Limbs*, that in tracing the manifold and progressive changes of the feet in the mammalian series, in those forms where it is normally composed of five digits, the middle is usually the largest; and this is the most constant one. The modifications in the hind foot, in reference to the number of digits, are, first, the reduction and then the removal, of the innermost one; then the corresponding reduction and removal of the outer one; next, of the second and fourth digits, until it is reduced to the middle digit, as in the horse.

The innermost toe, the first to dwindle and disappear in the brute series, is, in Man, developed to a maximum size, becoming emphatically the 'great toe,' one of the most essential characteristics of the human frame. It is made the powerful fulcrum for that lever of the second kind, which has its resistance in the tibio-astragalar joint, and the power applied to the projecting heelbone: the superincumbent weight is carried further forward upon the foot, by the more advanced position of the astragalus, than in the ape tribe; and the heel-bone is much stronger, and projects more backwards.

The arrangement of the powerfully-developed tarsal and metatarsal bones is such as to form, in Man, a bony arch, of which the two piers rest upon the proximal joint of the great toe and the end of the heel. Well-developed cuneiform bones combine with the cuboid to form a second arch, transverse to the first. There are no such modifications in the gorilla or orang, in which the arch, or rather the bend of the long and narrow sole, extends to the extreme end of the long and curved digits, indicating a capacity for grasping. Upon these two arches the superincumbent weight of man is solidly and sufficiently maintained, as upon a low dome, with this further advantage, that the different joints, cartilages, coverings, and synovial membranes, give a certain elasticity to the dome, so that in leaping, running, or dropping from a height, the jar is diffused and broken before it can be transmitted to affect the enormous brain-expanded cranium. The lower limbs in man are longer in proportion to the trunk than in any other known mammalian animal. The kangaroo might seem to be an exception, but if the hind limbs of the kangaroo are measured in relation to the trunk, they are shorter than in the human subject. In no animal is the femur so long in proportion to the leg as in man. In none does the tibia expand so much at its upper end. Here it presents two broad, shallow cavities, for the reception of the condyles of the femur. Of these condyles, in man only is the innermost longer than the outermost; so that the shaft of the bone inclines a little outwards to its upper end, and joins a 'neck' longer than in other animals, and set on at a very open angle. The weight of the body, received by the round heads of the thigh bones, is thus transferred to a broader base, and its support in the upright posture facilitated.

There is also the collateral advantage of giving more space to those powerful adductor muscles that assist in fixing the pelvis and trunk upon the hind limbs. With regard to the form of the pelvis, the iliac bones, compared with those in the gorilla, are short and broad : they are more bent forwards, the better to receive and sustain the abdominal viscera, and are more expanded behind to give adequate attachment to the powerful glutei muscles, which are developed to a maximum in the human species, in order to give a firm hold of the trunk upon the limbs, and a corresponding power of moving the limbs upon the trunk. The tuberosities of the ischium are rounded, not angular, and not inclined outwards, as in the gorilla and the rest of the ape tribe. The symphysis pubis is shorter than in the apes. The tail is reduced to three or four stunted vertebræ, anchylosed to form the bone called 'os coccygis.' The true vertebræ, as they are called in human anatomy, correspond in number with those of the chimpanzee and the orang, and in their divisions with the latter species, there being twelve thoracic, five lumbar, and seven cervical. This movable part of the column is distinguished by a beautiful series of sigmoid curves, convex forwards in the loins, concave in the back, and again slightly convex forwards in the neck. The cervical vertebræ, instead of having long spinous processes, have short processes, usually more or less bifurcated. The bodies of the true vertebræ increase in size from the upper dorsal to the last lumbar, which rests upon the base of the broad wedge-shaped sacrum, fixed obliquely between the sacro-iliac articulations. All these curves of the vertebral column, and the interposed elastic cushions, have relation to the libration of the head and upper limbs, and the diffusion and the prevention of the ill effects from shocks in many modes of locomotion which man, thus organised for an erect position, is capable of performing. The arms of man are brought into more symmetrical proportions with the lower limbs; and their bony framework shews all the perfections that have been superinduced upon it in the mammalian series, viz., a complete clavicle, the antibrachial bones so adjusted as to permit the rotary movements of pronation and supination, as well as of flexion and extension; manifesting those characters which adapt them for the manifold application of that most perfect and beautiful of prehensile instruments, the hand. The scapula is broad, with the glenoid articulation turned outwards; the clavicles are bent in a slight sigmoid flexure; the humerus exceeds in length the bones of the fore-arm. The carpal bones are eight in number. The thumb is developed far beyond any degree exhibited by the highest quadrumana, and is the most perfect opposing digit in the animal creation.

The skull is distinguished by the enormous expansion of the brain-case; by the restricted growth of the bones of the face, especially of the jaws, in relation to the small, equally-developed teeth; and by the early obliteration of the maxillo-intermaxillary suture. To balance the head upon the neck-bone, we find the condyles of the occiput brought forward almost to the centre of the base of the skull, resting upon the two cups of the atlas, so that there is but a slight tendency to incline forwards when the balancing action of the muscle ceases, as when the head nods during sleep, in an upright posture. Instead of the strongly developed occipital crest, we find a great development of true mastoid processes advanced nearer to the middle of the sides of the basis cranii, and of which there is only the rudiment in the gorilla. The upper convexity of the cranium is not interrupted by any sagittal or parietal cristæ. The departure from the archetype, in the human skull, is most conspicuous, in the vast expanse of the neural spines of the three chief cranial vertebræ, viz. occipital, parietal, and frontal.

'To what extent,' it may next be asked, 'does man depart from the typical character of his species?' With regard to the kind and amount of variety in mankind, we find, propagable and characteristic of race, a difference of stature, a difference in regard to colour of skin, difference in both colour and texture of the hair, and certain differences in the osseous framework.

As to stature, the Bushmen of South Africa and the natives of Lapland exhibit the extreme of diminution, ranging from four to five feet. Some of the Germanic races and the Patagonian Indians exhibit the opposite extreme, ranging from six to seven feet. The medium size prevails generally throughout the races of mankind.

With reference to the characteristics of colour, which are extreme, we have now opportunities of knowing how much that character is the result of the influence of climate. We know it more particularly by that most valuable mode of testing such influences which we derive from the peculiarity of the Jewish race. For 1800 years that race has been dispersed in different latitudes and climates, and they have preserved themselves distinct from intermixture with other races of mankind. There are some Jews still lingering in the valleys of the Jordan, having been oppressed by the successive conquerors of Syria for ages,—a low race of people, and described by trustworthy travellers as being as black as any of the Ethiopian races. Others of the Jewish people, participating in European civilization, and dwelling in the northern nations, shew instances of the light complexion, the blue eyes, and light hair of the Scandinavian families. The condition of the Hebrews, since their dispersion, has not been such as to admit of much admixture by the proselytism of household slaves. We are thus led to account for the differences in colour, by the influence of climate, without having to refer them to original or specific distinctions.

As to the difference in size in mankind, it is slight in comparison with what we observe in the races of the domestic dog, where the extremes of size are much greater than can be found in any races of the human species.

With reference to the modifications of the bony structure, as characteristic of the races of mankind, they are almost confined to the pelvis and the cranium. In the pelvis the difference is a slight, yet apparently a constant one. The pelvis of the adult negro may sometimes be distinguished from that of the European by the greater proportional length and less proportional breadth of the iliac bones; but how trifling is this difference compared with that marked distinction in the pelvis which the gorilla and orangoutang present!

With regard to the cranial differences, I have selected for comparison three extreme specimens of skulls characteristic of race: one of an aboriginal of Van Diemen's Land (the lowest of the Melanian or dark-coloured family), a well-marked Mongolian, and a well-formed European skull. The differences are chiefly these. In the low, uneducated, uncivilised races, the brain is rather smaller than in the higher, more civilised, and more educated races; consequently the cranium rises and expands in a less degree. Concomitant with this contraction of the brain-case is a greater projection of the fore part of the face; whether it may be from a longer exercise of the practice of suckling, or a more habitual application of the teeth in the premaxillary part of the jaw, and in the corresponding part of the lower jaw, in biting and gnawing tough, raw, uncooked substances,—the anterior alveolar part of the jaws does project more in those lower races; but still to an insignificant degree compared with the prominence of that part of the skull in the large apes. And while alluding to them, I may again advert to the distinction between them and the lowest of the

H

human races, which is afforded by the pre-maxillary bone, already referred to. In the young orang-utan, even when the change of dentition has begun, the suture between that bone and the maxillary is present; and it is not until the large canine teeth are developed, that the stimulus of the vascular system, in the concomitant expansion and growth of the alveoli, tends to obliterate the suture. In the young chimpanzee, the maxillary suture disappears earlier, at least on the facial surface of the upper jaw. In the human subject those traces disappear still earlier, and in regard to the exterior alveolar plates, the inter-maxillary and maxillary bones are connate. But there may be always traced in the human fœtus the indications of the palatal and nasal portions of the maxillo-intermaxillary suture, of which the poet Goethe was the first to appreciate the full significance.

In the Mongolian skull there is a peculiar development of the cheek-bones, giving great breadth and flatness to the face, a broad cranium, with a low forehead, and often with the sides sloping away from the median sagittal tract, something like a roof; whereas, in the European, there is combined, with greater capacity of the cranium, a more regular and beautiful oval form, a loftier and more expanded brow, a minor prominence of the malars, and a less projection of the upper and lower jaws. All these characteristics necessarily occasion slight differences in the facial angle. On a comparison of the basis cranii, the strictly bimanous characteristics in the position of the foramen magnum and occipital condyles, and of the zygomatic arches, are as well displayed in the lowest as in the highest varieties of the human species.

With regard to the value to be assigned to the above defined distinctions of race:—in consequence of not any of these differences being equivalent to those characteristics of the skeleton, or other parts of the frame, upon which specific differences are founded by naturalists in reference to the rest of the animal creation, I have come to the conclusion that Man forms one species, and that these differences are but indicative of varieties. As to the number of these varieties :—from the very well marked and natural character of the species, just as in the case of the similarly natural and circumscribed class of birds, scarcely any two ethnologists agree as to the number of the divisions, or as to the characters upon which those varieties are to be defined and circumscribed. In the subdivision of the class of birds, the ornithological systems vary from two orders to thirty orders; so with man there are classifications of races varying from *thirty* to the *three* predominant ones which Blumenbach first clearly pointed out,—the Ethiopian, the Mongolian, and the Caucasian or Indo-European. These varieties merge into one another by easy gradations. The Malay and the Polynesian link the Mongolian and the Indian varieties; and the Indian is linked by the Esquimaux again to the Mongolian. The inhabitants of the Andaman Islands, New Caledonia, New Guinea, and Australia, in a minor degree seem to fill up the hiatus between the Malayan and the Ethiopian varieties; and in no case can a well marked definite line be drawn between the physical characteristics of allied varieties, these merging more or less gradationally the one into the other.

In considering the import and value of the osteological differences between the gorilla—the most anthropoid of all known brutes—and man, in reference to the hypothesis of the origination of species of animals by gradual transmutation of specific characters, and that in the ascending direction:—it may be admitted that the skeleton is modifiable to a certain extent by the action of the muscles to which it is subservient, and that in domesticated races the size of the animal may be brought to deviate in both directions from the specific standard. By the development of the processes, ridges, and crests, and also by the general proportions of the bones themselves, especially those of the limbs, the human anatomist judges of the muscular power of the individual to whom a skeleton under comparison has appertained.

The influence of muscular actions in the growth of bone is more strikingly displayed in the change of form which the cranium of the young carnivore or the sternum of the young bird undergoes in the progress of maturity; not more so, however, than is manifested in the progress of the development of the cranium of the gorilla itself, which results in a change of character so great, as almost to be called a metamorphosis.

In some of the races of the domestic dog, the tendency to the development of parietal and occipital cristæ is lost, and the cranial dome continues smooth and round from one generation of the smaller spaniel, or dwarf pug, e.g. to another; while, in the large deer-hound, those bony cristæ are as strongly developed as in the wolf. Such modifications, however, are unaccompanied by any change in the connexions, that is, in the disposition of the sutures, of the cranial bones; they are due chiefly to arrests of development, to retention of more or less of the characters of immaturity : even the large proportional size of the brain in the smaller varieties of house-dog is in a great degree due to the rapid acquisition by the cerebral organ of its specific size, agreeably with the general law of its development, but which is attended in the varieties cited by an arrest of the general growth of the body, as well as of the particular developments of the skull in relation to the muscles of the jaws.

No species of animal has been subject to such decisive experiments, continued through so many generations, as to the influence of different degrees of exercise of the muscular system, difference in regard to food, association with man, and the concomitant stimulus to the development of intelligence, as the dog; and no domestic animal manifests so great a range of variety in regard to general size, to the colour and character of the hair, and to the form of the head, as it is affected by different proportions of the cranium and face, and by the intermuscular crests superadded to the cranial parietes. Yet, under the extremest mask of variety so superinduced, the naturalist detects in the dental formula and in the construction of the cranium the unmistakeable generic and specific characters of the *Canis familiaris.*

This and every other analogy applicable to the present question justifies the conclusion that the range of variety allotted to the gorilla, chimpanzee, and orang-utan, under the operation of external circumstances favourable to their higher development, would be restricted to differences of size, of colour, and other characters of the hair, and of the shape of the head, in so far as this is influenced by the arrest of general growth after the acquisition by the brain of its mature proportions, and by the development, or otherwise, of processes, crests, and ridges for the attachment of muscles. The most striking deviations from the form of the human cranium which that part presents in the great orangs and chimpanzees result from the latter acknowledged modifiable characters, and might be similarly produced; but not every deviation from the cranial structure of man, nor any of the important ones upon which the naturalist relies for the determination of the genera *Troglodytes* and *Pithecus*, have such an origin or dependent relation. The gorilla, indeed, differs specifically from both the orang and man in one cranial character, which no difference of diet, habit, or muscular exertion can be conceived to affect.

The prominent superorbital ridge, for example, is not the consequence or concomitant of muscular development; there are no muscles attached to it that could have excited its growth. It is a characteristic of the cranium of the genus *Troglodytes* from the time of birth to extreme old age; by the prominent superorbital ridge, for example, the skull of the young gorilla or chimpanzee with deciduous teeth may be distinguished at a glance from the skull of an orang at the same immature age; the genus *Pithecus*, Geoffr., being as well recognised by the absence, as the genus *Troglodytes* is by the presence, of this character. We have no grounds, from observation or experiment, to believe the absence or the presence of a prominent superorbital ridge to be a modifiable character, or one to be gained or lost through the operations of external causes, inducing particular habits through successive generations of a species. It may be concluded, therefore, that such feeble indication of the superorbital ridge, aided by the expansion of the frontal sinuses, as exists in man, is as much a specific peculiarity of the human skull, in the present comparison, as the exaggeration of this ridge is characteristic of the chimpanzees and its suppression of the orangs.

The equable length of the human teeth, the concomitant absence of any diastema or break in the series, and of any sexual difference in the development of particular teeth, are to be viewed by the light of actual knowledge, as being primitive and unalterable specific peculiarities of man.

Teeth, at least such as consist of the ordinary dentine of mammals, are not organised so as to be influenced in their growth by the action of neighbouring muscles; pressure upon their bony sockets may affect the direction of their growth after they are protruded, but not the specific proportions and forms of the crowns of teeth of limited and determinate growth. The crown of the great canine tooth of the male *Troglodytes gorilla* began to be calcified when its diet was precisely the same as in the female, when both sexes derived their sustenance from the mother's milk. Its growth proceeded and was almost completed before the sexual development had advanced so as to establish those differences of habits, of force, of muscular exercise, which afterwards characterise the two sexes. The whole crown of the great canine is, in fact, calcified before it cuts the gum or displaces its small deciduous predecessor; the weapon is prepared prior to the development of the forces by which it is to be wielded; it is therefore a structure fore-ordained, a predetermined character of the great ape, by which that creature is made physically superior to man; and one can as little conceive the development of the canine tooth to be a result of external stimulus, or as being influenced by the muscular actions, as the development of the stomach, or of any internal gland.

The two external divergent fangs of the premolar teeth, and the slighter modifications of the crowns of the molars and premolars, appear likewise from the actual results of observation to be equally predetermined and non-modifiable characters.

No known cause of change productive of varieties of mammalian species could operate in altering the size, the shape, or the connexions of the premaxillary bones, which so remarkably distinguish the *Troglodytes gorilla*, not from man only, but from all other anthropoid apes. We know as little the conditions which protract the period of the obliteration of the sutures of the premaxillary bones in the *Tr. gorilla* beyond the period at which they disappear in the *Tr. niger*, as we do those that cause them to disappear in man earlier than they do even in the smaller species of chimpanzee.

There is not, in fact, any other character than those founded upon the developments of bone for the attachment of muscles, which is known to be subject to change through the operation of external causes; nine-tenths, therefore, of the differences, especially those very striking ones manifested by the pelvis and pelvic extremities, which I have cited in the memoirs on the subject, published in the *Zoological Transactions*, as distinguishing the gorilla and chimpanzee from the human species, must stand in contravention of the hypothesis of transmutation and progressive development, until the supporters of that hypothesis are enabled to adduce the facts and cases which demonstrate the conditions of the modifications of such characters.

If the consideration of the cranial and dental characters of the *Troglodytes gorilla* has led legitimately to the conclusion that it is specifically distinct from the *Troglodytes niger*, the hiatus is still greater that divides it from the human species, between the extremest varieties of which there is no osteological and dental distinction which can be compared to that manifested by the shorter premaxillaries and larger incisors of the *Troglodytes niger* as compared with the *Tr. gorilla*.

The analogy which the establishment of the second and more formidable species of chimpanzee in Africa has brought to light between the representation of the genus *Troglodytes* in that continent, and that of the genus *Pithecus* in the great islands of the Indian Archipelago, is very close and interesting. As the *Troglodytes gorilla* parallels the *Pithecus Wurmbii*, so the *Troglodytes niger* parallels the *Pithecus morio,* and an unexpected illustration has thus been gained of the soundness of the interpretation of the specific distinction of that smaller and more anthropoid orang.

It is not without interest to observe, that as the generic forms of the *Quadrumana* approach the *Bimanous* order, they are represented by fewer species. The gibbons (*Hylobates*) scarcely number more than half-a-dozen species; the orangs (*Pithecus*) have but two species, or at most three; the chimpanzees (*Troglodytes*) are represented by two species.

The unity of the human species is demonstrated by the constancy of those osteological and dental characters to which the attention is more particularly directed in the investigation of the corresponding characters in the higher *Quadrumana.*

Man is the sole species of his genus, the sole representative of his order and subclass.

Thus I trust has been furnished the confutation of the notion of a transformation of the ape into man, which appears from a favourite old author to have been entertained by some in his day.

"And of a truth, vile epicurism and sensuality will make the soul of man so degenerate and blind, that he will not only be content to slide into brutish immorality, but please himself in this very opinion that he is a real brute already, an ape, satyre or baboon; and that the best of men are no better, saving that civilising of them and industrious education has made them appear in a more refined shape, and long inculcate precepts have been mistaken for connate principles of honesty and natural knowledge; otherwise there be no indispensable grounds of religion and virtue, but what has happened to be taken up by *over-ruling* custom. Which things, I dare say, are as easily confutable, as any conclusion in mathematics is demonstrable. But as many as are thus sottish, let them enjoy their own wildness and ignorance; it is sufficient for a good man that he is conscious unto himself that he is more nobly descended, better bred and born, and more skilfully taught by the purged faculties of his own minde [1]."

[1] Henry More's *Conjectura Cabbalistica,* fol. (1662)—p. 175.

Printed in the United States
By Bookmasters